PHYSICS RESEARCH AND TECHNOLOGY

ELECTRON GAS

AN OVERVIEW

PHYSICS RESEARCH AND TECHNOLOGY

Additional books and e-books in this series can be found on Nova's website under the Series tab.

PHYSICS RESEARCH AND TECHNOLOGY

ELECTRON GAS

AN OVERVIEW

TATA ANTONIA
EDITOR

Copyright © 2019 by Nova Science Publishers, Inc.

All rights reserved. No part of this book may be reproduced, stored in a retrieval system or transmitted in any form or by any means: electronic, electrostatic, magnetic, tape, mechanical photocopying, recording or otherwise without the written permission of the Publisher.

We have partnered with Copyright Clearance Center to make it easy for you to obtain permissions to reuse content from this publication. Simply navigate to this publication's page on Nova's website and locate the "Get Permission" button below the title description. This button is linked directly to the title's permission page on copyright.com. Alternatively, you can visit copyright.com and search by title, ISBN, or ISSN.

For further questions about using the service on copyright.com, please contact:
Copyright Clearance Center
Phone: +1-(978) 750-8400   Fax: +1-(978) 750-4470   E-mail: info@copyright.com.

NOTICE TO THE READER

The Publisher has taken reasonable care in the preparation of this book, but makes no expressed or implied warranty of any kind and assumes no responsibility for any errors or omissions. No liability is assumed for incidental or consequential damages in connection with or arising out of information contained in this book. The Publisher shall not be liable for any special, consequential, or exemplary damages resulting, in whole or in part, from the readers' use of, or reliance upon, this material. Any parts of this book based on government reports are so indicated and copyright is claimed for those parts to the extent applicable to compilations of such works.

Independent verification should be sought for any data, advice or recommendations contained in this book. In addition, no responsibility is assumed by the Publisher for any injury and/or damage to persons or property arising from any methods, products, instructions, ideas or otherwise contained in this publication.

This publication is designed to provide accurate and authoritative information with regard to the subject matter covered herein. It is sold with the clear understanding that the Publisher is not engaged in rendering legal or any other professional services. If legal or any other expert assistance is required, the services of a competent person should be sought. FROM A DECLARATION OF PARTICIPANTS JOINTLY ADOPTED BY A COMMITTEE OF THE AMERICAN BAR ASSOCIATION AND A COMMITTEE OF PUBLISHERS.

Additional color graphics may be available in the e-book version of this book.

Library of Congress Cataloging-in-Publication Data

ISBN: 978-1-53616-428-2

Published by Nova Science Publishers, Inc. † New York

CONTENTS

Preface		vii
Chapter 1	Electron Gas on the Surface of a Nanotube: Thermodynamics, Dynamic Conductivity, and Collective Phenomena *A. M. Ermolaev and G. I. Rashba*	1
Chapter 2	Study of the Transport of Charge Carriers in Materials with Degenerate Electron Gas *Vilius Palenskis and Evaras Žitkevičius*	123
Chapter 3	Enhanced Output Power of Ingan-Based Light-Emitting Diodes with AlGaN/GaN Two-Dimensional Electron Gas Structure *Jae-Hoon Lee*	187
Index		209
Related Nova Publications		215

PREFACE

In Electron Gas: An Overview, the results of theoretical studies of the thermodynamic, kinetic, and high-frequency properties of the electron gas on the surface of a nanotube in a magnetic field in the presence of a longitudinal superlattice are presented.

Following this, an interpretation of the basic transport characteristics of metals, superconductors in the normal state, and very strongly doped semiconductors with degenerate electron gas is presented. An application of the effective density of randomly moving electrons allows for an explanation of the conductivity of metals, and the electron transport characteristics such as the average diffusion coefficient, the average mobility, the mean free path, and the electron scattering mechanisms in a wide temperature range.

Finally, the authors demonstrate high-performance InGaN-based light-emitting diodes with tunneling-junction-induced 2-D electron gas at an AlGaN/GaN heterostructure, which is inserted in the middle of the P++-GaN contact layer of a conventional LED structure.

Chapter 1 - The results of theoretical studies of the thermodynamic, kinetic, and high-frequency properties of the electron gas on the surface of a nanotube in a magnetic field in the presence of a longitudinal superlattice are presented. Nano-dimensions of the motion area lead to energy quantization.

Its multiply connected structure in the presence of a magnetic field leads to effects that are derived from the Aharonov-Bohm effect. It is shown that the curvature of a nanotube, even in the absence of a magnetic field, causes new macroscopic oscillation effects such as de Haas-van Alphen oscillations, which are associated with the quantization of the transverse electron motion energy and with the root peculiarities of the density of electron states on the nanotube surface. Thermodynamic potentials and heat capacity of the electron gas on the tube are calculated in the gas approximation. The Kubo formula for the conductivity tensor of the electron gas on the nanotube surface is obtained. The Landau damping regions of electromagnetic waves on a tube are determined and the beats are theoretically predicted on the graph of the dependence of conductivity on tube parameters. In the hydrodynamic approximation the plasma waves on the surface of a semiconductor nanotube with a superlattice are considered. It is shown that optical and acoustic plasmons can propagate along a tube with one kind of carrier. An exact expression for the polarization operator of a degenerate electron gas on the surface of a nanotube was obtained. The shape and size of Landau damping area of plasma waves on a tube in the entire Brillouin zone are calculated. The conditions for the resonance absorption of plasmons on a tube by electrons are found. Electron spin waves on the surface of a semiconductor nanotube with a superlattice in a magnetic field are studied. The spectra and areas of collisionless damping of these waves are found. The authors have shown that the spin wave damping is absent in these areas if the tubes with a degenerate electron gas have small radius.

Chapter 2 - This study is addressed to the stochastic description of the effective density of the randomly moving (RM) electrons in metals and other materials with degenerate electron gas. It is written in the accessible form for researchers, engineers and students without an extensive background of quantum mechanics of solid-state physics. The chapter begins with the interpretation of the basic transport characteristics of the metals, superconductors in the normal state, and very strongly doped semiconductors with degenerate electron gas. An application of the effective density of RM electrons leads one simply to explain the conductivity of metals, and the electron transport characteristics such as the average diffusion coefficient, the average mobility, the mean free path, and the electron scattering mechanisms in very wide temperature range. The generalized expressions for basic electron transport characteristics, which are valid for materials both with non-degenerate and degenerate electron gas, are presented. It is well known that electrons obey the Pauli principle and they are described by the Fermi-Dirac statistics, and using the total density of free valence electrons for estimation of transport characteristics of electrons in materials with degenerate electron gas is unacceptable with respect to Fermi-Dirac statistics, because all these characteristics are determined by RM electrons near the Fermi level energy. An application of the classical statistics leads to colossal errors in the estimation of transport characteristics of electrons in materials with degenerate electron gas. It is shown that the Einstein's relation between the diffusion coefficient and drift mobility of RM electrons is held at any level of degeneracy of electron gas. The presented general expressions are applied for estimation of the carrier transport characteristics in the superconductor $YBa_2Cu_3O_{7-x}$ in the normal state, especially for description of the Hall-effect. It is shown that drift mobility of electrons in materials with degenerate electron gas can be tens or hundred times larger than the Hall mobility. The calculation results of the resistivity and other transport characteristics for elemental metals in temperature range from 1 K to 900 K are represented and compared with the experimental data.

Chapter 3 – The authors demonstrate high-performance InGaN-based light-emitting diodes (LEDs) with tunneling-junction-induced 2-D electron gas (2DEG) at an AlGaN/GaN heterostructure, which is inserted in the middle of the P^{++}-GaN contact layer of a conventional LED structure. The output power of a LED with a 2DEG insertion layer shows 20% enhancement compared to that of a conventional LED at 350 mA. This enhancement in output power for the LED with a 2DEG insertion layer could be attributed to both enhanced hole-injection efficiency and lateral current spreading by the presence of 2DEG at the AlGaN/GaN heterostructure.

In: Electron Gas: An Overview
Editor: Tata Antonia

ISBN: 978-1-53616-428-2
© 2019 Nova Science Publishers, Inc.

Chapter 1

ELECTRON GAS ON THE SURFACE OF A NANOTUBE: THERMODYNAMICS, DYNAMIC CONDUCTIVITY, AND COLLECTIVE PHENOMENA

A. M. Ermolaev and G. I. Rashba[*]
Theoretical Physics Department
named after academician I. M. Lifshits,
V. N. Karazin Kharkiv National University,
Kharkov, Ukraine

ABSTRACT

The results of theoretical studies of the thermodynamic, kinetic, and high-frequency properties of the electron gas on the surface of a nanotube in a magnetic field in the presence of a longitudinal superlattice are presented. Nano-dimensions of the motion area lead to energy quantization.

---

[*] Corresponding Author's E-mail: georgiy.i.rashba@gmail.com.

Its multiply connected structure in the presence of a magnetic field leads to effects that are derived from the Aharonov-Bohm effect. It is shown that the curvature of a nanotube, even in the absence of a magnetic field, causes new macroscopic oscillation effects such as de Haas-van Alphen oscillations, which are associated with the quantization of the transverse electron motion energy and with the root peculiarities of the density of electron states on the nanotube surface. Thermodynamic potentials and heat capacity of the electron gas on the tube are calculated in the gas approximation. The Kubo formula for the conductivity tensor of the electron gas on the nanotube surface is obtained. The Landau damping regions of electromagnetic waves on a tube are determined and the beats are theoretically predicted on the graph of the dependence of conductivity on tube parameters. In the hydrodynamic approximation the plasma waves on the surface of a semiconductor nanotube with a superlattice are considered. It is shown that optical and acoustic plasmons can propagate along a tube with one kind of carrier. An exact expression for the polarization operator of a degenerate electron gas on the surface of a nanotube was obtained. The shape and size of Landau damping area of plasma waves on a tube in the entire Brillouin zone are calculated. The conditions for the resonance absorption of plasmons on a tube by electrons are found. Electron spin waves on the surface of a semiconductor nanotube with a superlattice in a magnetic field are studied. The spectra and areas of collisionless damping of these waves are found. We have shown that the spin wave damping is absent in these areas if the tubes with a degenerate electron gas have small radius.

Keywords: nanotubes, superlattice, magnetic field, thermodynamic functions, dynamic conductivity, plasma waves, electron spin waves.

1. INTRODUCTION

Almost thirty years have passed since the discovery of carbon nanotubes by Iijima [1]. However, the interest in these nanosystems is so great that in recent years a new direction in physics and technology has emerged – carbon nanomaterial science. Many articles and reviews have appeared in the world of scientific literature (see, for example, [2-4]), in which the properties of nanotubes are studied. Interest in them is due to the fact that nanotubes are functional elements of many scientific instruments and technical devices. They are used in nanotransistors, nanodiodes and nanoscale diodes, displays, sensors, during transportation and storage of toxic substances. They are interesting to physicists because nanotubes are dielectric, semiconductor, metal, so the methods developed to study these systems are transferred to nanotubes and other electron nanosystems on curved surfaces [4]. To study their properties, it is necessary to synthesize the methods of quantum mechanics, statistics and kinetics and Riemannian geometry. A new parameter appearing in theory (the curvature of the structure) contributes to enriching the picture of phenomena in nanosystems increasing the ways to control their properties. In electronic systems on curved surfaces, effects have already been discovered that have no analogue in systems with flat geometry. These include effects of hybridization of size and magnetic quantization of the motion of conduction electrons, modification of the electron Hamiltonian [4], specific resonances in the scattering of electrons in carbon nanotubes [5] and quantum wires [6] by impurity atoms.

In Section 2 we present the results of calculations of some thermodynamic quantities of an electron gas on the surface of a cylindrical nanotube. We use the method of contour integration developed by Rumer [7, 8]. The degenerated electron gas and non-degenerated gas are considered.

The interest of scientists in semiconducting heterostructures, quantum wells [9, 10], nanosystems on curved surfaces [11] increases. Physicists have learned how to make nanotubes with a diameter of several angstrom under laboratory conditions. The authors of the work [12] obtained ropes and films of single-wall carbon nanotubes oriented along the magnetic field. The carbon nanotubes in the presence of a magnetic field were used for producing membranes in Ref. [13]. Magnetization of the polyimide nanocomposite samples containing carbon nanotubes was measured as a function of temperature and magnetic field intensity in Ref. [14]. Various effects have been discovered in nanotubes, which include the effects of spatial and magnetic quantization of conduction electron motion in a magnetic field, modification of the Hamiltonian of a system [4, 15], unusual behavior of conductance [16] and magnetic response [17-19], peculiarities of screening of the Coulomb interaction of electrons [20, 21], specific resonances of electron scattering in carbon nanotubes [5] and quantum wires [6] by impurity atoms. The conduction electron energy spectrum in the semiconductor nanotubes has a band nature. A small electron density near the band edge permits to use the effective mass approximation. This approximation allows describing the properties of such systems qualitatively and often also quantitatively. The thermodynamic functions of electron gas on a nanotube surface have been studied in literature [17-19, 22-25]. In Ref. [22] the geometrical effects in ideal quantum gases of electrons, photons and phonons in confined space were considered. In Ref. [23] a thermodynamic analysis of the boron-nitride nanotubes nucleation on the catalysts surface was performed. In Refs. [24, 25] the chemical potential, energy, pressure and the work function of an electronic gas on a conducting carbon nanotube surface under zero temperature are calculated. Within the framework of the Hartree-Fock approximation the contact electron-electron interaction is taken into account. Analytical form of the work function of carbon nanotubes was derived in the paper [25]. At large radii of nanotubes the limit to the work function of graphene was done. However, in Refs. [22-25], the magnetic fields were not included, the nondegenerate electron gas and the oscillations of thermodynamic values were not studied. In Subsections 2.1, 2.2 we present the results of calculations of such thermodynamic functions of the degenerate and nondegenerate electron gas on the semiconductor cylindrical nanotube surface in a longitudinal magnetic field as chemical potential, internal energy, grand potential, heat capacity, entropy, and spin magnetization. We employ the effective mass approximation and the Laplace transform for the calculation of the density of states [7, 26, 27].

Solution to the problem of energy spectrum recovery of metals found by I. M. Lifshits and his collaborators Refs. [28-30] turned out to be useful in the studies of other systems. Among those are low-dimensional nanosystems, organic conductors and layered systems with metallic conductivity character. Utilization of this concept is also appropriate in studying the properties of an electron gas found at the surface of a nanotube with a superlattice. The interest in systems with superlattices is due to a number of reasons. These systems are the functional elements of many modern technical devices. Improvement of the method for molecular-beam epitaxy allows creating such systems in the laboratory. Existence in the theory of systems with superlattices of additional parameters (amplitude and modulating potential period) increases the number of ways to control its properties. The logic behind the development of solid state physics is such that currently the objects of investigations are not only three dimensional systems with superlattices [31, 32], but also low-dimensional systems. Method of a molecular-beam epitaxy allows creation of synthetic periodic structure not only in three-dimensional systems and in a two-dimensional electron gas, but also on the surface of carbon [1] and semiconducting [4] nanotubes. Superlattice at the surface of a carbon nanotube has been previously studied [33]. Authors estimated orbital magnetization of the electron gas at the surface of the nanotube with the superlattice in a magnetic field parallel to the axes of the tube and the superlattice. Using the model suggested by the authors [33], we calculated the heat capacity of the degenerate electron gas. Just as in the abovementioned article [33], here we suggest that the bare spectrum of electron energy is singleband and parabolic.

The authors of the work [34] had solved the Maxwell's equation taking into account the retardation for the conducting cylinder dipped in a dielectric medium. They considered time dispersion of the dielectric permittivity of conductor and dielectric. Also they had solved the dispersion equation for the wave spectrum, obtained new branches in the spectrum of surface polaritons. The article [35] deals with polaritons in the magnetic wire with uniaxial anisotropy dipped into a dielectric medium. Wire's dielectric and magnetic permittivity tensor frequency dependence also was studied. Polaritons spectra were found. Constant, parallel to the magnetic wire axis, external magnetic field effect was studied in article [36]. In this article was shown how wave characteristics allow one to obtain data about the material structure. A necessary of semiconductor nanotubes waveguide characteristics researching is caused by growing interest to these systems. Plasma and magnetoplasma waves propagation on the nanotube surface were studied in Refs. [37-44]. They focused on cylindrical geometry systems calculations. These calculations were made by means of matter dielectric permittivity simple models. The problem is that authors of Refs. [37-44] have not used dielectric permittivity and conductivity tensors exact expressions. Also they have not taken into account the time and spatial dispersion. However, these exact formulas may significantly influence the tube waveguide characteristics expressions.

In Subsections 3.1 - 3.3, we present conductivity tensor components calculation and their wave vector and frequency dependencies of the following system: electron gas on the nanotube surface that is affected by a parallel to the tube axis external magnetic field.

Interest in carbon [1-3] and semiconductor [4, 45, 46] nanotubes is caused by their unique properties, namely, high strength and conductivity, as well as magnetic, waveguide and optical properties. These systems are prepared by rolling up a graphene sheet (or two-dimensional heterostructure) into a tube. Depending on the rolling up manner, the tube has metallic, semiconductor, or dielectric properties. Modern technologies allow creating not only nanotubes but nanotubes with superlattices. Along with flat superlattices [47-54], also ones with cylindrical symmetry exist [55]. They are of radial and longitudinal types [55, 56]. The radial superlattice is a set of coaxial cylinders, while the longitudinal one looks like a set of coaxial rings of the same radius. The tubes with longitudinal superlattice are prepared by lithographic methods by embedding the fullerenes into the tube. In such a system, there exists the periodic potential acting upon electrons moving along the tube. In the electron energy spectrum the minibands appear. The electron density of states has root singularities at the miniband boundaries. In connection with increased interest in currents within the cylindrical conductors, the authors of Ref. [57] have calculated the longitudinal conductivity for solid and hollow cylinders without superlattice in magnetic field and considered quantum electromagnetic waves in such systems. Exact expressions for all the components of the conductivity tensor for degenerate and nondegenerate electron gas on the nanotube surface without superlattice are presented in Subsection 3.1. It is worth to be clarified how the superlattice affects this tensor. In the Subsections 3.2, 3.3 the dynamic conductivity tensor components were calculated based on the model of effective mass for the nanotube with longitudinal superlattice in magnetic field. The superlattice axis and the magnetic intension vector were considered to be parallel to the tube axis. A review of articles on collective excitations of electron gas on the tube for the period up to 2014 is contained in the collection of articles [58].

In the framework of the hydrodynamic approach, the plasma waves on the surface of a nanotube with a longitudinal superlattice in Subsections 4.1 are considered. Not only longitudinal electron current but also transversal one has been taken into consideration. It has been shown that both optical and acoustical plasmons could propagate along the tube with one sort of carrier. Within the framework of the model electron energy spectrum on the nanotube surface with a superlattice in a magnetic field, an exact expression for the polarization operator of a degenerate electron gas was obtained in Subsection 4.2.

The shape and size of the plasma waves Landau damping regions on the tube throughout the Brillouin zone were calculated. The influence on these areas of the position of the Fermi level in the miniband was considered. The conditions for the resonance absorption of plasmon on the tube by electrons were found. The limiting transition towards the nanotube without superlattice was performed. Electron spin waves on the surface of a semiconductor nanotube with a superlattice in a magnetic field have been considered in the Subsection 4.3. The spin-wave spectra and regions of collisionless wave damping have been found. It has been shown that the spin waves do not exhibit damping on small-radius tubes with a degenerate electron gas.

The thermodynamic functions of the electron gas on the surface of a nanotube are calculated in papers [59-61]. The dynamic conductivity tensor of the electron gas on the tube is calculated in papers [62-64]. The papers [65-67] published after 2013, theoretically investigated the properties of plasma and spin waves on a tube. In General conclusion, the results of the papers [59-67] are discussed.

2. THERMODYNAMIC FUNCTIONS OF AN ELECTRON GAS ON A TUBE

2.1. Thermodynamic Functions of an Electron Gas on a Nanotube Surface in the Absence of a Magnetic Field

The electron gas on the surface of the nanotube is formed by valence electrons that are not involved in the formation of bonds. The energy spectrum of an electron has a zonal character [2-4]. However, accounting for a small number of electrons at the edge of the zone, one can use the effective mass approximation. Then, using the symmetry of a cylindrical nanotube, we will characterize the state of an electron on it with an azimuthal quantum number $m = 0, \pm 1, \pm 2, \ldots$ and projection k wave vector on the tube axis.

The wave function of the stationary state of electron $|mk\rangle$ has the form:

$$\psi_{mk}(\varphi, z) = \frac{e^{im\varphi}}{\sqrt{2\pi}} \frac{e^{ikz}}{\sqrt{L}},$$

(2.1.1)

where φ is the polar angle, z is the cylinder axis, L is the tube length. The electron energy in the state (2.1.1) is equal to

$$\varepsilon_{mk} = \varepsilon_0 m^2 + \frac{\hbar^2 k^2}{2m_*}.$$

(2.1.2)

Here m_* is the effective electron mass, $\varepsilon_0 = \hbar^2 / 2m_* a^2$ is the rotational quantum, a is the tube radius. The spectrum (2.1.2) consists of a set of one-dimensional subzones whose boundaries $\varepsilon_0 m^2$ are not equidistant.

Density of electron states

$$v(\varepsilon) = \sum_{mk\sigma} \delta(\varepsilon - \varepsilon_{mk}) = \frac{L}{\pi\hbar} \sqrt{2m_*} \sum_m (\varepsilon - \varepsilon_0 m^2)^{-\frac{1}{2}} \qquad (2.1.3)$$

has a root peculiarities at the boundaries of subbands in the spectrum (2.1.3). Here σ is the spin quantum number, and summation is performed on m, for which the radical expression is positive ($|m| \leq \left[\sqrt{\varepsilon/\varepsilon_0}\right]$).

Using the Laplace transform, we represent the density of states (2.1.3) in the form of a contour integral [7, 8]:

$$v(\varepsilon) = \frac{1}{2\pi i} \int\limits_{c-i\infty}^{c+i\infty} du\, z(u)\, e^{u\varepsilon}, \qquad (2.1.4)$$

where

$$z(u) = \sum_{mk\sigma} \exp(-u\varepsilon_{mk}) \qquad (2.1.5)$$

is the single-particle statistical sum and the constant $c > 0$ selected such that all the singularities of the integrand lie to the left of the integration contour in (2.1.4). We transform the summation over m included in (2.1.5) using the formula [7, 8, 68, 69]:

$$\sum_{m=-\infty}^{\infty} \exp\left[-x(m+v)^2\right] = \sqrt{\frac{\pi}{x}}\, \Theta_3\left(v, e^{-\frac{\pi^2}{x}}\right) =$$

$$= \sqrt{\frac{\pi}{x}} \sum_{l=-\infty}^{\infty} \exp\left(-\frac{\pi^2 l^2}{x}\right) \cos(2\pi l v), \qquad (2.1.6)$$

where

$$\Theta_3(\upsilon,q) = \sum_{m=-\infty}^{\infty} q^{m^2} \cos(2\pi m\upsilon) \quad (|q|<1)$$

is the theta-function, $x > 0$.

The result is

$$z(u) = \frac{L}{\hbar u}\sqrt{\frac{2m_*}{\varepsilon_0}} \sum_{l=-\infty}^{\infty} \exp\left(-\frac{\pi^2 l^2}{u\varepsilon_0}\right). \tag{2.1.7}$$

Substituting (2.1.7) into (2.1.4) and using the integral representation of the Bessel function J_n [70]

$$J_n(\alpha z) = \frac{z^n}{2\pi i} \int_{c-i\infty}^{c+i\infty} du\, u^{-n-1} \exp\left[\frac{\alpha}{2}\left(u - \frac{z^2}{u}\right)\right] \tag{2.1.8}$$

($c, \alpha > 0$, $n > -1$), we find

$$v(\varepsilon) = \frac{2m_* aL}{\hbar^2}\left[1 + 2\sum_{l=1}^{\infty} J_0\left(2\pi l\sqrt{\frac{\varepsilon}{\varepsilon_0}}\right)\right]. \tag{2.1.9}$$

With $\varepsilon \gg \varepsilon_0$ this implies

$$v(\varepsilon) = \frac{2m_* aL}{\hbar^2}\left[1 + \frac{2}{\pi}\left(\frac{\varepsilon_0}{\varepsilon}\right)^{1/4}\sum_{l=1}^{\infty}\frac{1}{\sqrt{l}}\cos\left(2\pi l\sqrt{\frac{\varepsilon}{\varepsilon_0}} - \frac{\pi}{4}\right)\right]. \tag{2.1.10}$$

The density of states (2.1.10) oscillates with change of $\sqrt{\varepsilon}$ around the density of states value

$$\frac{2m_*aL}{\hbar^2} \tag{2.1.11}$$

of two-dimensional electron gas over the area $S = 2\pi aL$. Period of oscillation of function $\nu(\varepsilon)$ (considered in dependence on $\sqrt{\varepsilon}$) equals to $\hbar/\sqrt{2m_*}a$. Relative amplitude of order oscillations is $(\varepsilon_0/\varepsilon)^{1/4}$. It is small at $\varepsilon_0 \ll \varepsilon$.

If the number of electrons N on the surface of the nanotube is fixed, the chemical potential μ of electron gas can be obtained from the formula [71]

$$N = \int_0^\infty d\varepsilon \nu(\varepsilon) f(\varepsilon), \tag{2.1.12}$$

where $f(\varepsilon)$ is the Fermi distribution function. Substituting (2.1.9) into (2.1.12) and performing integration by parts, we obtain the equation for μ

$$N = \frac{L}{\hbar}\sqrt{\frac{2m_*}{\varepsilon_0}}\int_0^\infty d\varepsilon\, \varepsilon \left(-\frac{df}{d\varepsilon}\right)\left[1 + \frac{2}{\pi}\sqrt{\frac{\varepsilon_0}{\varepsilon}}\sum_{l=1}^\infty \frac{1}{l}J_1\left(2\pi l\sqrt{\frac{\varepsilon}{\varepsilon_0}}\right)\right]. \tag{2.1.13}$$

Chemical potential oscillates with change n with a period $\hbar/\sqrt{2m_*}a$ around value

$$\mu_0 = \hbar n\sqrt{\frac{\varepsilon_0}{2m_*}} \tag{2.1.14}$$

of chemical potential of a two-dimensional electron gas over the area $S = 2\pi aL$. Here $n = N/L$ is the electron linear density.

In the ultraquantum limit when only the subzone $m = 0$ is filled at $k_B T \ll \mu$ (T is the temperature, k_B is the Boltzmann's constant) from formula (2.1.13) we obtained the main contribution to the chemical potential:

$$\mu = \mu_0 \left[1 + \frac{\pi^2}{12} \left(\frac{k_B T}{\mu_0} \right)^2 \right], \qquad (2.1.15)$$

where

$$\mu_0 = \frac{\pi^2 \hbar^2 n^2}{8 m_*}. \qquad (2.1.16)$$

The internal energy of the electron gas on the surface of the nanotube is equal to:

$$E = \frac{L}{\hbar} \sqrt{\frac{2 m_*}{\varepsilon_0}} \int_0^\infty d\varepsilon \, \varepsilon \, f(\varepsilon) \left[1 + 2 \sum_{l=1}^\infty J_0 \left(2\pi l \sqrt{\frac{\varepsilon}{\varepsilon_0}} \right) \right]. \qquad (2.1.17)$$

If the parameter μ/ε_0 is not too large, the main terms of the expansion of internal energy (2.1.17) in powers of $k_B T/\mu$ take the form

$$E = E_0 + \frac{1}{2} CT, \qquad (2.1.18)$$

where

$$E_0 = \frac{L}{\hbar}\sqrt{\frac{m_*}{2\varepsilon_0}}\mu_0^2 \times$$

$$\times\left\{1+\frac{4\varepsilon_0}{\pi^2\mu_0}\sum_{l=1}^{\infty}\frac{1}{l^2}\left[J_2\left(2\pi l\sqrt{\frac{\mu_0}{\varepsilon_0}}\right)-\pi l\sqrt{\frac{\mu_0}{\varepsilon_0}}J_3\left(2\pi l\sqrt{\frac{\mu_0}{\varepsilon_0}}\right)\right]\right\} \tag{2.1.19}$$

is the ground state energy,

$$C = \frac{L}{\hbar}\sqrt{\frac{2m_*}{\varepsilon_0}}\frac{\pi^2}{3}k_B^2 T \times$$

$$\times\left\{1+2\sum_{l=1}^{\infty}\left[J_0\left(2\pi l\sqrt{\frac{\mu_0}{\varepsilon_0}}\right)-\pi l\sqrt{\frac{\mu_0}{\varepsilon_0}}J_1\left(2\pi l\sqrt{\frac{\mu_0}{\varepsilon_0}}\right)\right]\right\} \tag{2.1.20}$$

is the heat capacity. In these formulas μ_0 is chemical potential (2.1.14) at $T=0$.

In the ultraquantum limit, from formula (2.1.17) we obtain

$$E = E_0\left[1+\frac{\pi^2}{4}\left(\frac{k_B T}{\mu_0}\right)^2\right], C = \frac{\pi k_B^2 L}{3\hbar}\sqrt{\frac{2m_*}{\mu_0}}T, \tag{2.1.21}$$

where

$$E_0 = \frac{2L}{3\pi\hbar}\sqrt{2m_*}\mu_0^{3/2}$$

and chemical potential μ_0 is equal to (2.1.16). From formula (2.1.16) it follows that the ultraquantum limit is reached at linear electron densities satisfying the inequality $n < (2/\pi a)$. This inequality can be satisfied in semiconductor nanotubes with a small number of conduction electrons.

Grand potential Ω of electron gas calculated by the formula [71]

$$\Omega = -\frac{1}{\beta}\int_0^\infty d\varepsilon\, \nu(\varepsilon)\ln\left(1+e^{\beta(\mu-\varepsilon)}\right), \qquad (2.1.22)$$

where β is the reverse temperature. Substituting here (2.1.9) and performing twice integration by parts, we obtain

$$\Omega = -\frac{L\varepsilon_0}{\hbar}\sqrt{\frac{m_*\varepsilon_0}{2}}\times$$

$$\times\int_0^\infty d\varepsilon\left(-\frac{df}{d\varepsilon}\right)\left(\frac{\varepsilon}{\varepsilon_0}\right)^2\left[1+\frac{4\varepsilon_0}{\pi^2\varepsilon}\sum_{l=1}^\infty\frac{1}{l^2}J_2\left(2\pi l\sqrt{\frac{\varepsilon}{\varepsilon_0}}\right)\right]. \qquad (2.1.23)$$

With $k_B T \ll \mu$ hence the main contribution to the entropy of the electron gas has the form:

$$S = \frac{L}{\hbar}\sqrt{\frac{2m_*}{\varepsilon_0}}\frac{\pi^2}{3}k_B^2 T. \qquad (2.1.24)$$

As in the three-dimensional electron gas, this expression coincides with the main contribution to the heat capacity (2.1.20).

From formulas (2.1.13), (2.1.17), (2.1.19), (2.1.20), (2.1.23) it can be seen that the thermodynamic quantities of the electron gas on the surface of the nanotube oscillate with a change of $\sqrt{\mu}$ with a period $\hbar/\sqrt{2m_*a}$. Measurement of period allows to obtaine the effective electron mass if the tube radius is known. These oscillations are similar to the de Haas-van Alphen oscillations of the magnetization of an electron gas with a change of the magnetic field. They are caused by a jump-like change in the density of states (2.1.9) when the boundaries of the subbands of the spectrum (2.1.2) cross the Fermi level. The latter can be changed by the method developed in the case of a two-dimensional electron gas. The amplitude of oscillations depends on the temperature and radius of the tube.

To calculate the thermodynamic functions of a nondegenerate electron gas on the surface of a nanotube, we use the Boltzmann distribution function

$$f(\varepsilon) = \exp\big(\beta(\mu - \varepsilon)\big). \tag{2.1.25}$$

The method of contour integration applied above allows us to obtain exact expressions for the chemical potential and the internal energy of the electron gas:

$$\mu = \frac{1}{\beta} \ln \left[n\hbar \sqrt{\frac{\varepsilon_0}{2m_*}} \beta \left(1 + 2 \sum_{l=1}^{\infty} e^{-\frac{\pi^2 l^2}{\beta \varepsilon_0}} \right)^{-1} \right], \tag{2.1.26}$$

$$E = \frac{L}{\hbar} \sqrt{\frac{2m_*}{\varepsilon_0}} \frac{1}{\beta^2} e^{\beta\mu} \left[1 + 2 \sum_{l=1}^{\infty} \left(1 - \frac{\pi^2 l^2}{\beta \varepsilon_0} \right) e^{-\frac{\pi^2 l^2}{\beta \varepsilon_0}} \right]. \tag{2.1.27}$$

The integral

$$\int_0^{\infty} dx\, e^{-ax^2} J_1(bx) = \frac{1}{b} \left[1 - \exp\left(-\frac{b^2}{4a} \right) \right], \qquad (a > 0).$$

[69] was used to derive these formulas. Asymptotics of the sum entering into (2.1.26) have the form:

$$\sum_{l=-\infty}^{\infty} e^{-l^2 x} = \begin{cases} \sqrt{\dfrac{\pi}{x}}\left(1 + 2e^{-\dfrac{\pi^2}{x}}\right), x \ll 1, \\ \\ 1 + 2e^{-x}, x \gg 1. \end{cases} \qquad (2.1.28)$$

If $\beta\varepsilon_0 \ll 1$ then the quantization of the circular motion of electrons can be neglected. Then, replacing the sums with integrals, we obtain the chemical potential of a non-degenerate two-dimensional electron gas in the area $S = 2\pi a L$:

$$\mu = \frac{1}{\beta}\ln\left(\frac{\pi\hbar^2 N\beta}{m_* S}\right). \qquad (2.1.29)$$

The internal energy of the Boltzmann electron gas on the surface of a nanotube can be represented as:

$$E = \frac{L}{\hbar}\sqrt{\frac{m_*}{2\pi}}\frac{1}{\beta^{3/2}}e^{\beta\mu} \times$$

$$\times \sum_{m=-\infty}^{\infty} \exp\left(-\beta\varepsilon_0 m^2\right)\left(1 + 2\beta\varepsilon_0 m^2\right) = \frac{Nk_B T}{2}\left(1 + 2\frac{\varepsilon_0}{k_B T}\left\langle m^2\right\rangle\right), \qquad (2.1.30)$$

where

$$\left\langle m^p\right\rangle = \frac{\displaystyle\sum_m m^p e^{-\beta\varepsilon_0 m^2}}{\displaystyle\sum_m e^{-\beta\varepsilon_0 m^2}}. \qquad (2.1.31)$$

The first term at the right-hand side of formula (2.1.30) is the average energy of electron motion along the tube. This motion is not quantized, so

Electron Gas on the Surface of a Nanotube

its energy $E_\parallel = Nk_BT/2$ coincides with that obtained on the basis of the theorem on the uniform distribution of energy among the degrees of freedom. The second term in (2.1.30) is equal to the average energy of the circular motion of electrons. If $\beta\varepsilon_0 \ll 1$, sums in (2.1.30) and (2.1.31) can be replaced by integrals. As a result, the expression for the energy of circular motion $E_\perp = Nk_BT/2$ also consistent with the uniform distribution energy theorem.

The heat capacity of a nondegenerate electron gas is:

$$C = \frac{Nk_B}{2}\left[1 + 2\left(\frac{\varepsilon_0}{k_BT}\right)^2\left(\langle m^4\rangle - \langle m^2\rangle^2\right)\right]. \tag{2.1.32}$$

Its asymptotics are equal:

$$C = \begin{cases} \dfrac{Nk_B}{2}\left[1 + 4\left(\dfrac{\varepsilon_0}{k_BT}\right)^2\exp\left(-\dfrac{\varepsilon_0}{k_BT}\right)\right], & k_BT \ll \varepsilon_0, \\ Nk_BT, & k_BT \gg \varepsilon_0. \end{cases} \tag{2.1.33}$$

As expected, the contribution of the circular motion of electrons to the heat capacity is exponentially small at low temperatures. This behavior is a consequence of the existence of a gap ε_0 in the spectrum of a circular motion. At a temperature of about ε_0/k_B the temperature dependence of heat capacity (2.1.32) has a maximum similar to the maximum rotational heat capacity of molecules [71]. At values $m_* = 0.07m_0$ (m_0 is the free electron mass) and $a = 10^{-6}\,cm$ typical for semiconductor nanostructures, this temperature is equal $\varepsilon_0/k_B = 64K$.

2.2. Thermodynamic Functions of Electron Gas on the Semiconductor Nanotube Surface in a Magnetic Field

The convenient quantum numbers that characterize the state of conduction electrons on the semiconductor cylindrical nanotube surface in a longitudinal magnetic field include $m = 0, \pm 1, \ldots$ is the azimuthal quantum number; k is the electron wave vector projection on the cylinder axis z; $\sigma = \pm 1$ is the spin quantum number. The electron energy in the state $|mk\sigma\rangle$ equals to

$$\varepsilon_{mk\sigma} = \varepsilon_0 \left(m + \frac{\Phi}{\Phi_0} \right)^2 + \frac{k^2}{2m_*} + \sigma\mu_B B, \qquad (2.2.1)$$

where m_* is the electron effective mass, $\varepsilon_0 = \left(2m_* a^2 \right)^{-1}$ is the rotational quantum, a is the tube radius, $\Phi = \pi a^2 B$ is the magnetic flux B in the tube cross-section, $\Phi_0 = 2\pi c / e$ is the flux quantum, e is the electron charge value, c is the speed of light, and μ_B is the spin magnetic moment of an electron. The Planck constant in (2.2.1) and hereinafter is assumed to be unity. The spectrum (2.2.1) is a set of one-dimensional subbands whose boundaries $\varepsilon_{m0\sigma}$ are non-equidistant.

Let us represent the density of electronic states

$$\nu(\varepsilon) = \sum_{mk\sigma} \delta\left(\varepsilon - \varepsilon_{mk\sigma} \right)$$

with the spectrum (2.2.1) as a contour integral [7, 8, 26, 27]:

$$v(\varepsilon) = \frac{1}{2\pi i} \int_{\alpha-i\infty}^{\alpha+i\infty} du \, z(u) e^{u\varepsilon}, \tag{2.2.2}$$

where

$$z(u) = \sum_{mk\sigma} e^{-u\varepsilon_{mk\sigma}} \tag{2.2.3}$$

is the one-particle statistical sum. The constant $\alpha > 0$ is chosen that all the peculiarities of the integrand in Eq. (2.2.2) should be located to left from the integration contour.

To calculate the sum over m in Eq. (2.2.2) the following formula is used [69]:

$$\sum_{m=-\infty}^{+\infty} \exp\left[-x(m+\upsilon)^2\right] = \sqrt{\frac{\pi}{x}} \sum_{l=-\infty}^{+\infty} \exp\left(-\frac{\pi^2 l^2}{x}\right) \cos(2\pi l\upsilon), \qquad x > 0.$$

Thus the sum (2.2.3) equals to

$$z(u) = \frac{L}{u} \sqrt{\frac{2m_*}{\varepsilon_0}} \, \mathrm{ch}\left(u\mu_B B\right) \sum_{l=-\infty}^{+\infty} \exp\left(-\frac{\pi^2 l^2}{u\varepsilon_0}\right) \cos\left(2\pi l \frac{\Phi}{\Phi_0}\right), \tag{2.2.4}$$

where L is the semiconductor tube length. By substituting Eq. (2.2.4) into Eq. (2.2.2), we obtain a contour integral that is calculated precisely using the integral representation of the Bessel function [70]:

$$J_n(pz) = \frac{z^n}{2\pi i} \int_{\alpha-i\infty}^{\alpha+i\infty} \frac{du}{u^{n+1}} \exp\left[\frac{p}{2}\left(u - \frac{z^2}{u}\right)\right] \qquad (\alpha, p > 0, \ n > -1) \cdot$$

As the result, the density of states is equal to

$$v(\varepsilon) = L\sqrt{\frac{2m_*}{\varepsilon_0}}\left[1 + \sum_{l=1}^{\infty}\cos\left(2\pi l\frac{\Phi}{\Phi_0}\right)\sum_{\sigma}J_0\left(2\pi l\sqrt{\frac{\varepsilon-\varepsilon_\sigma}{\varepsilon_0}}\right)\right], \quad (2.2.5)$$

where $\varepsilon_\sigma = \sigma\mu_B B$.

The function (2.2.5) contains a monotonic summand

$$v_0 = 2m_*aL \qquad (2.2.6)$$

and oscillating summands. The expression (2.2.6) equals to the two-dimensional electron gas density of states over the area

$$S = 2\pi aL. \qquad (2.2.7)$$

This system is obtained by cutting the tube along its generatrix and transforming it into a strip with the area (2.2.7). The oscillating summands in (2.2.5) are caused by the superposition of the de Haas-van Alphen and the Aharonov-Bohm type oscillations. The former are described by multiplier J_0 and the latter by multiplier $\cos\left(2\pi l\frac{\Phi}{\Phi_0}\right)$. The de Haas-van Alphen type oscillations are caused by the variable ε crossing the subband boundaries in spectrum (2.2.1). The density of states has root peculiarities. These oscillations survive also in the absence of a magnetic field. Shown in Figure 1 is a plot of the first harmonic of the function $\frac{v}{v_0}$ depending on $\frac{\varepsilon}{\varepsilon_0}$ in the absence of a magnetic field. The Aharonov-Bohm type oscillations are linked to variation of the magnetic flux Φ passing via the tube cross section. The period of these oscillations equals to the flux quantum Φ_0.

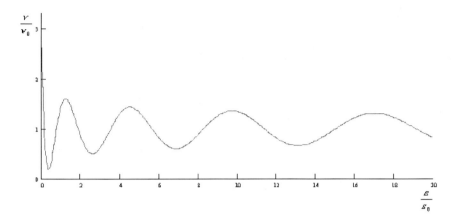

Figure 1. The first harmonic of the density of states (2.2.5) as a function of $\varepsilon/\varepsilon_0$ at $B=0$.

If the Bessel function argument is large, we use the formula (2.2.5) to get

$$v(\varepsilon) = v_0 \times$$

$$\times \left[1 + \frac{2}{\pi} \sum_\sigma \left(\frac{\varepsilon_0}{\varepsilon - \varepsilon_\sigma} \right)^{1/4} \sum_{l=1}^{\infty} \frac{1}{\sqrt{l}} \cos\left(2\pi l \sqrt{\frac{\varepsilon - \varepsilon_\sigma}{\varepsilon_0}} - \frac{\pi}{4} \right) \cos\left(2\pi l \frac{\Phi}{\Phi_0} \right) \right], \quad (2.2.8)$$

where $\varepsilon - \varepsilon_\sigma \gg \varepsilon_0$. With increasing of $(\varepsilon - \varepsilon_\sigma)^{1/2}$ the density of states (2.2.8) oscillates around the value (2.2.6) with a period of

$$\tau = \left(\sqrt{2m_* a} \right)^{-1}. \quad (2.2.9)$$

In the case of the constant number of electrons N on the semiconductor tube surface, the chemical potential μ of electron gas can be found from the following formula:

$$N = \int_0^\infty d\varepsilon\, v(\varepsilon) f(\varepsilon) = L\sqrt{\frac{m_*}{2\varepsilon_0}} \sum_\sigma \int_{\varepsilon_\sigma}^\infty d\varepsilon \left(-\frac{df}{d\varepsilon}\right)(\varepsilon - \varepsilon_\sigma) \times$$

$$\times \left[1 + \frac{2}{\pi}\sqrt{\frac{\varepsilon_0}{\varepsilon - \varepsilon_\sigma}} \sum_{l=1}^\infty \frac{1}{l} J_1\left(2\pi l \sqrt{\frac{\varepsilon - \varepsilon_\sigma}{\varepsilon_0}}\right) \cos\left(2\pi l \frac{\Phi}{\Phi_0}\right)\right], \tag{2.2.10}$$

where $f(\varepsilon)$ is the Fermi function. At $\mu \gg \varepsilon_0$ and $\beta\mu \gg 1$ (β is the inverse temperature) the equation takes the form

$$N = \sqrt{\frac{2m_*}{\varepsilon_0}}\mu L\left[1 + \frac{1}{\beta\mu}\sum_\sigma\left(\frac{\varepsilon_0}{\mu_\sigma}\right)^{1/4} \times \right.$$

$$\left. \times \sum_{l=1}^\infty \frac{1}{\sqrt{l}}\cos\left(2\pi l\frac{\Phi}{\Phi_0}\right)\sin\left(2\pi l\sqrt{\frac{\mu_\sigma}{\varepsilon_0}} - \frac{\pi}{4}\right)\mathrm{sh}^{-1}\left(\frac{\pi^2 l}{\beta(\varepsilon_0\mu_\sigma)^{1/2}}\right)\right],$$

$$\mu_\sigma = \mu - \varepsilon_\sigma.$$

The monotonic summand in the right hand side of equation (2.2.10) links the Fermi energy of degenerate electron gas with the linear electron density $n = N/L$:

$$\mu_0 = \sqrt{\frac{\varepsilon_0}{2m_*}}n. \tag{2.2.11}$$

The chemical potential undergoes the de Haas-van Alphen type oscillations around the value (2.2.11). The relative amplitude of these oscillations at $\mu_0 \gg \varepsilon_0, \mu_B B$ is about $\left(\varepsilon_0/\mu_0\right)^{3/4}$. They are caused by the passage of Fermi energy via the subband boundaries $\varepsilon_{m0\sigma}$. The period

Electron Gas on the Surface of a Nanotube

of these oscillations with variation of $\sqrt{\mu_0}$ is equal to (2.2.9). When the magnetic flux Φ changes, the chemical potential oscillates with the period Φ_0.

In the ultra quantum limit, when one subband only is filled $m = -M$, $\sigma = -1$, the degenerate gas chemical potential equals to:

$$\mu = \varepsilon_{\min} + \mu_q \left[1 + \frac{\pi^2}{12} \left(\frac{T}{\mu_q} \right)^2 \right],$$

where

$$\varepsilon_{\min} = \varepsilon_0 \eta^2 - \mu_B B , \quad \mu_q = \frac{\pi^2 n^2}{2 m_*},$$

T is the temperature in units of erg, M and η are the integer and the fractional parts of Φ / Φ_0. The ultra quantum limit regime at $B = 0$ can be achieved in semiconductor nanotubes with small electron density at $na < 2/\pi$.

The internal electron energy on the tube surface equals to:

$$E = L \sqrt{\frac{m_*}{2\varepsilon_0}} \times$$

$$\times \sum_\sigma \int_{\varepsilon_\sigma}^\infty d\varepsilon \, \varepsilon f(\varepsilon) \left[1 + 2 \sum_{l=1}^\infty J_0 \left(2\pi l \sqrt{\frac{\varepsilon - \varepsilon_\sigma}{\varepsilon_0}} \right) \cos \left(2\pi l \frac{\Phi}{\Phi_0} \right) \right].$$

(2.2.12)

The monotone contribution into the degenerate electron gas energy is

$$E_0 = L\sqrt{\frac{m_*}{2\varepsilon_0}}\mu_0^2\left[1-\left(\frac{\mu_B B}{\mu_0}\right)^2\right].$$

The energy (2.2.12) oscillates when n and Φ change with the periods (2.2.9) and Φ_0 respectively. Oscillating energy contribution (2.2.12) at $\mu \gg \varepsilon_0$ and $\mu \gg T$ equals to

$$E_{osc} = \sqrt{2m_*}LT \times$$

$$\times\sum_{l=1}^{\infty}\frac{1}{\sqrt{l}}\cos\left(2\pi l\frac{\Phi}{\Phi_0}\right)\sum_{\sigma}\sqrt{\mu_\sigma}\sin\left(2\pi l\sqrt{\frac{\mu_\sigma}{\varepsilon_0}}-\frac{\pi}{4}\right)\text{sh}^{-1}\left(\frac{\pi^2 lT}{\left(\varepsilon_0\mu_\sigma\right)^{\frac{1}{2}}}\right).$$

If the Bessel function argument is not too large, the electron gas heat capacity at $T \ll \mu_0$ can be derived from (2.2.12) using power expansion T/μ_0 [71,72]:

$$C = \frac{\pi^2}{3}L\sqrt{\frac{2m_*}{\varepsilon_0}}T\left\{1+\sum_{l=1}^{\infty}\sum_{\sigma}\left[\text{J}_0\left(2\pi l\sqrt{\frac{\mu_0-\varepsilon_\sigma}{\varepsilon_0}}\right)-\right.\right.$$

$$\left.\left.-\pi l\frac{\mu_0}{\varepsilon_0}\sqrt{\frac{\varepsilon_0}{\mu_0-\varepsilon_\sigma}}\text{J}_1\left(2\pi l\sqrt{\frac{\mu_0-\varepsilon_\sigma}{\varepsilon_0}}\right)\right]\cos\left(2\pi l\frac{\Phi}{\Phi_0}\right)\right\}.$$

The main contribution into the heat capacity is

$$C_0 = \frac{\pi^2}{3}L\sqrt{\frac{2m_*}{\varepsilon_0}}T. \qquad (2.2.13)$$

This value increases as aT grows. The plot of C/C_0 depending on Φ/Φ_0 at $\mu_0/\varepsilon_0 = 10$ is shown in Figure 2.

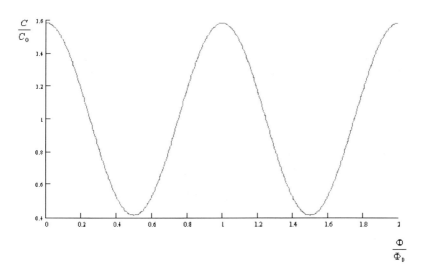

Figure 2. Plot of degenerate electron gas heat capacity against magnetic flux at $\mu_0/\varepsilon_0 = 10$.

In the ultra quantum limit at $T \ll \mu$, we obtain

$$E = \frac{L}{3\pi}\sqrt{2m_*}\mu_q^{3/2}\left[1 + 3\frac{\varepsilon_{\min}}{\mu_q} + \frac{\pi^2}{4}\left(\frac{T}{\mu_q}\right)^2\right],$$

$$C = \frac{\pi L}{3}\sqrt{\frac{m_*}{2\mu_q}}\,T.$$

In this case, the heat capacity does not depend on the tube radius.

With a fixed chemical potential and variable number of electrons at the tube surface, the grand electron gas potential at $\varepsilon_0 \ll \mu$, $T \ll \mu$ equals to:

$$\Omega = -\frac{L}{4}\sqrt{\frac{2m_*}{\varepsilon_0}}\sum_{\sigma}\mu_{\sigma}^2\left[1+\frac{\pi^2}{3}\left(\frac{T}{\mu_{\sigma}}\right)^2+\frac{4}{\pi}\left(\frac{\varepsilon_0}{\mu_{\sigma}}\right)^{5/4}\frac{T}{(\varepsilon_0\mu_{\sigma})^{1/2}}\times\right.$$
$$\left.\times\sum_{l=1}^{\infty}\frac{1}{l^{3/2}}\cos\left(2\pi l\frac{\Phi}{\Phi_0}\right)\cos\left(2\pi l\sqrt{\frac{\mu_{\sigma}}{\varepsilon_0}}-\frac{\pi}{4}\right)\mathrm{sh}^{-1}\left(\frac{\pi^2 l T}{(\varepsilon_0\mu_{\sigma})^{1/2}}\right)\right].$$

By differentiating this expression over T and B, we obtain entropy and magnetic moment of electron gas as functions of T, μ and B. The main contribution to entropy equals to

$$S = \frac{\pi^2 L}{3}\sqrt{\frac{2m_*}{\varepsilon_0}}T.$$

Numerically, the entropy is equal to the heat capacity (2.2.13), similarly to a three-dimensional case. Since the diamagnetic moment has been analyzed in the papers [17-19] without taking into account de Haas-van Alphen oscillation, we would confine ourselves to the monotonic part of the electron gas spin magnetization. Similarly to a three-dimensional case [71], it is proportional to the density of states:

$$M_S = \nu_0\mu_B^2 B.$$

This is the main contribution into the two-dimensional electron gas spin magnetization over the area (2.2.7) [73].

Electron Gas on the Surface of a Nanotube 27

At high temperature the Boltzmann distribution function is used for calculating thermodynamic quantities of electron gas on the semiconductor tube surface in a magnetic field:

$$f(\varepsilon) = \exp\left[\beta(\mu - \varepsilon)\right].$$

The chemical potential for constant electron number is found from the equation (2.2.10):

$$\mu(\beta, n) =$$

$$= \frac{1}{\beta} \ln \left\{ \beta n \sqrt{\frac{\varepsilon_0}{2m_*}} \left[\operatorname{ch}(\beta\mu_B B)\left(1 + 2\sum_{l=1}^{\infty} \exp\left(-\frac{\pi^2 l^2}{\beta\varepsilon_0}\right)\cos 2\pi l \frac{\Phi}{\Phi_0}\right)\right]^{-1}\right\} \qquad (2.2.14)$$

This function undergoes the Aharonov-Bohm type oscillations with the change of the flux Φ. The amplitude of these oscillations is small at $\beta\varepsilon_0 \ll 1$. The monotonic summand in Eq. (2.2.14) coincides with the chemical potential of the two-dimensional Boltzmann electron gas on the area (2.2.7). Shown in Figure 3 is the temperature dependence of nondegenerate electron gas chemical potential.

The nondegenerate electron gas internal energy equals to:

$$E = \frac{L}{\beta^2}\sqrt{\frac{2m_*}{\varepsilon_0}}\, e^{\beta\mu}\left\{\operatorname{ch}\beta\mu_B B - \beta\mu_B B \operatorname{sh}\beta\mu_B B + \right.$$

$$+ 2\sum_{l=1}^{\infty}\cos\left(2\pi l \frac{\Phi}{\Phi_0}\right)\exp\left(-\frac{\pi^2 l^2}{\beta\varepsilon_0}\right)\times \qquad (2.2.15)$$

$$\left.\times\left[\left(1 - \frac{\pi^2 l^2}{\beta\varepsilon_0}\right)\operatorname{ch}\beta\mu_B B - \beta\mu_B B \operatorname{sh}\beta\mu_B B\right]\right\}.$$

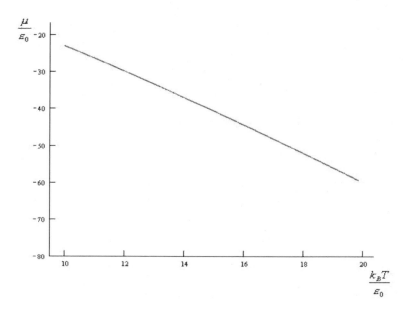

Figure 3. Plot of nondegenerate electron gas chemical potential (2.2.14) against temperature for $na = 1$, $\Phi/\Phi_0 = 2$ and $\beta\varepsilon_0 \ll 1$, $\beta\mu_B B \ll 1$.

This function is also a subject to the Aharonov-Bohm oscillations. The exact expressions (2.2.14) and (2.2.15) have been obtained by using integrals [69]:

$$\int_0^\infty dx\, e^{-px} J_0\left(q\sqrt{x}\right) = \frac{1}{p}\exp\left(-\frac{q^2}{4p}\right),$$

$$\int_0^\infty dx\, e^{-px^2} J_1(qp) = \frac{1}{q}\left[1 - \exp\left(-\frac{q^2}{4p}\right)\right]$$

$(\operatorname{Re} p > 0).$

In the absence of the magnetic field, the internal energy (2.2.15) may be represented as

$$E = \frac{N}{2\beta}\left(1 + 2\beta\varepsilon_0 \left\langle m^2 \right\rangle\right), \tag{2.2.16}$$

where

$$\left\langle m^l \right\rangle = \frac{\sum\limits_m m^l e^{-\beta\varepsilon_0 m^2}}{\sum\limits_m e^{-\beta\varepsilon_0 m^2}}.$$

The first summand in the right side of the formula (2.2.16) is caused by electron motion along the semiconductor tube which is not quantized. It follows immediately from the classical theorem of regular energy distribution over the degrees of freedom [71]. The second summand is connected to the quantization of electron circular motion.

We use Eq. (2.2.16) to obtain the Boltzmann gas heat capacity:

$$C = \frac{N}{2}\left[1 + 2(\beta\varepsilon_0)^2 \left(\left\langle m^4 \right\rangle - \left\langle m^2 \right\rangle^2\right)\right]. \tag{2.2.17}$$

Its asymptotics are as follows:

$$C = \begin{cases} \dfrac{N}{2}\left[1 + 4\beta\varepsilon_0 \exp\left(-\beta\varepsilon_0\right)\right], & \beta\varepsilon_0 \gg 1, \\ N, & \beta\varepsilon_0 \ll 1. \end{cases} \tag{2.2.18}$$

The contribution of electron circular motion into the heat capacity at $\beta\varepsilon_0 \gg 1$ is small. This results from the existence of the gap ε_0 in the circular motion excitation spectrum. At high temperatures, the expression (2.2.18) for heat capacity agrees with the theorem of regular distribution of energy over the degrees of freedom. The function of the heat capacity (2.2.17) against the temperature has a peak at a temperature of the order of

ε_0. It is similar to the peak of the molecular rotational heat capacity [71]. At the values of $m_* = 0.07m_0$ (m_0 is the free electron mass) and $a = 100$ angstrom which are typical for semiconductor nanotubes, this temperature equals to $\varepsilon_0 = 64K$.

2.3. Thermodynamic Functions of an Electron Gas at the Surface of a Nanotube with Its Superlattice in a Magnetic Field

Energy of an electron with an effective mass m_* at the surface of a cylindrical nanotube with its superlattice subject to a magnetic field parallel to its axes equals [31, 74]

$$\varepsilon_{mk\sigma} = \varepsilon_0 \left(m + \frac{\Phi}{\Phi_0} \right)^2 + \Delta(1 - \cos kd) + \sigma\mu_B B, \qquad (2.3.1)$$

where $m = 0, \pm 1, \ldots$ is the azimuthal quantum number, k is the projection of the electron wave vector on the axis z, $\sigma = \pm 1$ is the spin quantum number, $\varepsilon_0 = 1/2m_*a^2$ is the rotational quantum, a is the tube radius, $\Phi = \pi a^2 B$ is the magnetic induction flux B through the cross-section of the tube, $\Phi_0 = 2\pi c/e$ is the flux quantum [74], Δ and d is the amplitude and period of the modulating potential of the superlattice [31], μ_B is the spin magnetic moment of the electron. Quantum constant is assumed to be unity here and further. If $kd \ll 1$, the energy of longitudinal electron motion is simplified to $k^2/2m_*$, where $m_* = 1/\Delta d^2$

Electron Gas on the Surface of a Nanotube

. We suppose that in the described case effective mass of an electron for longitudinal and perpendicular motion is the same.

Electron energy state density with the spectrum (2.3.1) is calculated according to the formula

$$v(\varepsilon) = \sum_{mk\sigma} \delta(\varepsilon - \varepsilon_{mk\sigma}).$$

This equals

$$v(\varepsilon) = \frac{L}{\pi d} \sum_{m\sigma} \frac{\theta(\varepsilon - \varepsilon_{m\sigma})\theta(\varepsilon_{m\sigma} + 2\Delta - \varepsilon)}{\sqrt{(\varepsilon - \varepsilon_{m\sigma})(\varepsilon_{m\sigma} + 2\Delta - \varepsilon)}}. \tag{2.3.2}$$

Here $\varepsilon_{m\sigma} = \varepsilon_{m0\sigma}$, L is the tube length, θ is the Heaviside function. In the absence of a superlattice equation (2.3.1) represents a system of one-dimensional subbands with root singularities of state density at their boundaries $\varepsilon_{m\sigma}$. Modulating potential converts this spectrum to minibands 2Δ wide with boundaries $\varepsilon_{m\sigma}$ and $\varepsilon_{m\sigma} + 2\Delta$.

Flux ratio $\eta = \Phi/\Phi_0$ is included in (2.3.2) in the form of $m + \eta$. This allows limiting η to $0 \le \eta \le 1$. The order of miniband location depends on η. If $\eta < 1/2$ we have $\varepsilon_0 \eta^2 < \varepsilon_{-1} < \varepsilon_{+1} < \varepsilon_{-2} < \dots$ If $\eta > 1/2$ then $\varepsilon_{-1} < \varepsilon_0 \eta^2 < \varepsilon_{-2} < \dots$ Here and below spin level splitting will not be taken into consideration. At $\eta < 1/2$ lower miniband is within $\left[\varepsilon_0 \eta^2, \varepsilon_0 \eta^2 + 2\Delta\right]$ and the next is within $\left[\varepsilon_{-1}, \varepsilon_{-1} + 2\Delta\right]$. Energy gap between them is equal to $\varepsilon_0(1 - 2\eta) - 2\Delta$. Width of the k-th gap between $(k + 1)$-th and k-th minibands $(k = 1, 3, \dots)$ is equal to

$\varepsilon_0 k(1-2\eta) - 2\Delta$. Usually, in experiments with nanotubes of radius $a \sim (10^{-7} \sim 10^{-6})$ cm ($\varepsilon_0 \gg \Delta$), the relationship between fluxes in different fields is far less than unity, therefore minibands don't overlap. However, with an increase in tube radius, their overlap is inevitable.

Figure 4 shows dimensionless density of states $A = \pi v d\varepsilon_0 / 2L$ (2.3.2) in the two lower minibands of spectrum (2.3.1) as a function of $\varepsilon/\varepsilon_0$ for parameters $\eta = 0.1$, $\Delta/\varepsilon_0 = 0.1$ usually used in experiments [33,75].

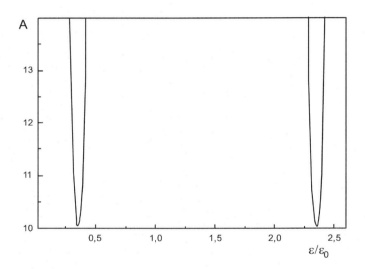

Figure 4. Electron energy state density (2.3.2) at the two lowest minibands of spectrum (2.3.1) for parameter values given in the text.

When $\varepsilon \gg \varepsilon_0$, the sum of m included in (2.3.2) can be substituted with an integral expression. As a result, the spectrum of the nanotubes becomes continuous, and electron state density is now equal to

$$
\nu(\varepsilon) = \begin{cases} \dfrac{4L}{\pi d \sqrt{2\Delta\varepsilon_0}} K\left(\sqrt{\dfrac{\varepsilon}{2\Delta}}\right), \varepsilon < 2\Delta, \\[4mm] \dfrac{4L}{\pi d \sqrt{\varepsilon\varepsilon_0}} K\left(\sqrt{\dfrac{2\Delta}{\varepsilon}}\right), \varepsilon > 2\Delta, \end{cases} \tag{2.3.3}
$$

where $K(k)$ is the complete elliptic integral of the first kind with modulus k [76]. Considering the abovementioned relationship between m_* with Δ and d, we are reassured that (2.3.3) represents the density of states of a two-dimensional electron gas with a one-dimensional superlattice in the absence of a magnetic field occupying a strip with area $S = 2\pi aL$. The energy of an electron in such a system equals [31, 77]

$$
\varepsilon(k_x, k_y) = \frac{k_x^2}{2m_*} + \Delta(1 - \cos k_y d).
$$

This system can be obtained by cutting the tube along its length and turning it inside out to form a surface. If $\varepsilon \ll 2\Delta$ from (2.3.3) the state density of a two-dimensional electron gas in the absence of a superlattice is obtained: $\nu_0 = m_* S / \pi$. At point $\varepsilon = 2\Delta$ the function (2.3.3) has a logarithmic form

$$
\nu(\varepsilon) = \frac{2L}{\pi d \sqrt{2\Delta\varepsilon_0}} \ln \frac{2\Delta}{|\varepsilon - 2\Delta|}.
$$

With energy increase the function decreases according to the law

$$
\nu(\varepsilon) = \frac{2L}{d \sqrt{\varepsilon\varepsilon_0}} \left(1 + \frac{\Delta}{2\varepsilon}\right), \qquad \varepsilon \gg 2\Delta.
$$

Poisson formula is used for the calculation of summation over m that included in expression (2.3.2) at $\varepsilon \gg \varepsilon_0$. Then $V = V_{mon} + V_{osc}$, where V_{mon} is the monotonic component of the density of states of (2.3.3) and V_{osc} is the oscillating component. The latter contains Fourier integral with a finite limits, where integrand has a root singularity at the limits of integration. Asymptote of the far Fourier component of this integral is known [78]. From it we obtained

$$v_{osc}(\varepsilon) = \frac{4L}{\pi d \varepsilon_0} \left(\frac{\varepsilon_0}{\varepsilon} \right)^{3/4} \left(\frac{\varepsilon}{2\Delta} \right)^{1/2} \sum_{l=1}^{\infty} \frac{1}{\sqrt{l}} \cos 2\pi l \frac{\Phi}{\Phi_0} \cos \left(2\pi l \sqrt{\frac{\varepsilon}{\varepsilon_0}} - \frac{\pi}{4} \right),$$
$$\varepsilon_0 \ll \varepsilon < 2\Delta, \qquad (2.3.4)$$

$$v_{osc}(\varepsilon) = \frac{4L}{\pi d \varepsilon_0} \left(\frac{\varepsilon_0}{\varepsilon} \right)^{3/4} \left(\frac{\varepsilon}{2\Delta} \right)^{1/2} \sum_{l=1}^{\infty} \frac{1}{\sqrt{l}} \cos 2\pi l \frac{\Phi}{\Phi_0} \times$$
$$\times \left[\cos \left(2\pi l \sqrt{\frac{\varepsilon}{\varepsilon_0}} - \frac{\pi}{4} \right) + \left(1 - \frac{2\Delta}{\varepsilon} \right)^{-1/4} \cos \left(2\pi l \sqrt{\frac{\varepsilon}{\varepsilon_0} \left(1 - \frac{2\Delta}{\varepsilon} \right)} + \frac{\pi}{4} \right) \right],$$
$$\varepsilon \gg \varepsilon_0, \varepsilon > 2\Delta.$$

Function (2.3.4) oscillates with the change in electron energy and magnetic flux Φ. The amplitude of oscillations decreases with increase in energy proportionally to $\varepsilon^{-1/4}$.

Using density of states (2.3.3) and (2.3.4) let us calculate the number of electrons N, their energy E, chemical potential μ and heat capacity C [71]. Let us consider degenerate gas at the surface of the nanotube.

In the case appropriate for nanotubes with a small radius, when at zero temperature electrons partially fill only the lower miniband, we obtained

$$N = \frac{4L}{\pi d}\arcsin\sqrt{\frac{\mu_0 - \varepsilon_-}{2\Delta}}$$

$$E = \frac{4L}{\pi d}\Delta\left[\left(1+\frac{\varepsilon_-}{\Delta}\right)\arcsin\sqrt{\frac{\mu_0 - \varepsilon_-}{2\Delta}} - \right. \tag{2.3.5}$$

$$\left. -\frac{1}{2\Delta}\sqrt{(\mu_0 - \varepsilon_-)(\varepsilon_- + 2\Delta - \mu_0)}\right]$$

Here $\varepsilon_- = \varepsilon_0\eta^2$ is the lower limit of spectrum (2.3.1), μ_0 is the Fermi energy. From (2.3.5) Fermi energy is found

$$\mu_0 = \varepsilon_- + 2\Delta\sin^2\frac{\pi dn}{4}.$$

The energy of a completely filled miniband is equal

$$E = \frac{2L\Delta}{d}\left(1+\frac{\varepsilon_-}{\Delta}\right),$$

where $n = N/L$ is linear electron density.

In order to obtain heat capacity of electron gases one must perform Sommerfeld expansion of the functions N and E of powers of T/μ, where T is the temperature (Boltzmann constant of 1 is assumed). This is possible if the chemical potential is located far from the features of state densities, i.e., the following inequalities must be met

$$T \ll \mu - \varepsilon_-, T \ll \varepsilon_+ - \mu, \tag{2.3.6}$$

where $\varepsilon_\pm$ are the upper and lower boundaries of the last partially filled miniband. Corrections on the order of T^2 in expansion of N and E are equal

$$N_T = \frac{\pi L T^2}{3d}\left(\mu_0 - \varepsilon_- + \Delta\right)\left[\left(\mu_0 - \varepsilon_-\right)\left(\varepsilon_+ - \mu_0\right)\right]^{-3/2},$$

$$E_T = \frac{\pi L T^2}{3d}\left[\mu_0\Delta - \varepsilon_-\left(\varepsilon_+ - \mu_0\right)\right]\left[\left(\mu_0 - \varepsilon_-\right)\left(\varepsilon_+ - \mu_0\right)\right]^{-3/2}.$$

If $T \ll \mu_0 - \varepsilon_- \ll 2\Delta$ then corrections in chemical potential and energy due to temperature are equal

$$\delta\mu = \frac{\pi^2 T^2}{12\left(\mu_0 - \varepsilon_-\right)},$$

$$\delta E = \frac{\pi L \Delta T^2}{3d\left(2\Delta\right)^{3/2}\sqrt{\mu_0 - \varepsilon_-}}. \tag{2.3.7}$$

δE takes into account a term present due to the dependence of chemical potential on temperature. From (2.3.7) we get monotonic component of the nanotube's heat capacity

$$C_{mon} = \frac{\pi L T}{3d\sqrt{2\Delta}\sqrt{\mu_0 - \varepsilon_-}}. \tag{2.3.8}$$

Using Eq. (2.3.3) heat capacity of an electron gas with a superlattice in the absence of a magnetic field at low temperatures can be obtained. If $\mu_0 < 2\Delta$ heat capacity equals

$$C = \frac{TS}{3d}\sqrt{\frac{m_*}{\Delta}}\left(1 - \frac{\mu_0}{2\Delta}\right)^{-1}\left[E\left(\sqrt{\frac{\mu_0}{2\Delta}}\right) + \left(1 - \frac{\mu_0}{2\Delta}\right)K\left(\sqrt{\frac{\mu_0}{2\Delta}}\right)\right], \tag{2.3.9}$$

where $E(k)$ is the complete elliptic integral of the second kind [76]. Coefficient at T in this formula is calculated precisely. If $\mu_0 \ll 2\Delta$, from (2.3.9) standard expression for the heat capacity of an electron gas without a superlattice is obtained:

$$C = \frac{\pi m_* TS}{3},$$

where density of states ν_0 is used. In accord with the Pauli principle heat capacity (2.3.8) and (2.3.9) is proportional to the temperature. However, proportionality coefficient is a complex function of the μ_0/Δ parameter.

Oscillating components N and E at conditions (2.3.6) and $\varepsilon_0 \ll \mu_0 < 2\Delta$ are equal

$$\binom{N_{osc}}{E_{osc}} = \frac{4(\varepsilon_0\mu_0)^{1/4} L}{\pi^2 d\sqrt{2\Delta}} \binom{1}{\mu_0} \times$$

$$\times \sum_{l=1}^{\infty} \frac{1}{l^{3/2}} \cos\left(2\pi l \frac{\Phi}{\Phi_0}\right) \sin\left(2\pi l \sqrt{\frac{\mu_0}{\varepsilon_0}} - \frac{\pi}{4}\right) \frac{\lambda_l}{sh\lambda_l},$$

(2.3.10)

where $\lambda_l = \pi^2 lT/(\varepsilon_0\mu_0)^{1/2}$. Functions (2.3.10) experience oscillations similar to de Haas-van Alphen and Aharonov-Bohm type oscillations with changes in $\mu_0^{1/2}$ related to electron density and magnetic flux Φ. The first are due to passage of root singularity of state density (2.3.2) at miniband boundaries through Fermi energy. This brings oscillations in consideration closer to de Haas-van Alphen type oscillations in a magnetic field [71]. However nonequidistance of energy levels of cross-sectional movement of electrons in the tube brings about $(\mu_0/\varepsilon_0)^{1/2}$ in phase with oscillations (2.3.10). These oscillations exist in absence of a magnetic field. Their period is equal to $\tau = 1/\sqrt{2m_* a}$. A measurement of the period allows one to obtain effective mass of an electron. Amplitude of oscillations decreases with an increase in temperature, as it does in the usual case of de Haas-van Alphen effect in a quantizing magnetic field [71]. Aharonov-Bohm type oscillations in normal and superconducting systems predicted in previous papers [74, 79] manifest in multiple phenomena [80-82].

From expression (2.3.10) let us obtain the oscillating term of heat capacity of a nanotube:

$$C_{osc} = \frac{4\mu_0 L}{d\sqrt{2\Delta}(\varepsilon_0\mu_0)^{1/4}} \sum_{l=1}^{\infty} \frac{1}{\sqrt{l}} \cos\left(2\pi l \frac{\Phi}{\Phi_0}\right) \times$$

$$\times \sin\left(2\pi l \sqrt{\frac{\mu_0}{\varepsilon_0}} - \frac{\pi}{4}\right) \frac{1}{\sh\lambda_1}(1 - \lambda_1 \cth\lambda_1).$$

(2.3.11)

With an increase in temperature monotonic component of heat capacity (2.3.8) exceeds the oscillating component (2.3.11) if $T > \mu_0 (\mu_0/\varepsilon_0)^{1/4}$.

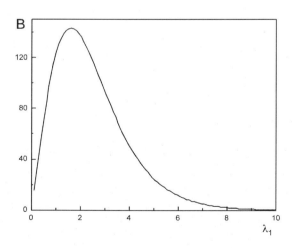

Figure 5. Temperature dependence of the amplitude of the oscillating component of heat capacity (2.3.11) for parameter values given in the text.

Figure 5 illustrates dependence of the amplitude of the main harmonic of the oscillating component of heat capacity (2.3.11)

$$B = \frac{4}{\sqrt{2}}\frac{L}{d}\left(\frac{\varepsilon_0}{\Delta}\right)^{1/2}\left(\frac{\mu_0}{\varepsilon_0}\right)^{3/4}\frac{\lambda_1 cth\lambda_1 - 1}{sh\lambda_1}$$

on the temperature when $\Phi/\Phi_0 = 0.1$ for the values of GaAs parameters that are usually used in experiments [75]: $m_* = 0.07m_0$ (where m_0 is the mass of a free electron), $a = 10^{-6} cm$, $\mu_0/\varepsilon_0 = 10$, $L = 10\mu m$, $\Delta = 1\ meV$, $d = 3500\overset{0}{A}$. Amplitude of B reaches its maximum value at temperature $T_m \propto (\varepsilon_0\mu_0)^{1/2}$ (α denotes proportionality).

At an assigned number of electrons chemical potential of a nondegenerate electron gas can be determined from equation

$$N = \sum_{mk\sigma} \exp\left[\beta(\mu - \varepsilon_{mk\sigma})\right], \qquad (2.3.12)$$

where β is the reverse temperature. Sums included in this expression are determined precisely. For estimating the sum by m the following formula is used [69]

$$\sum_{m=-\infty}^{\infty} \exp\left[-x(m+\upsilon)^2\right] = \sqrt{\frac{\pi}{x}}\sum_{l=-\infty}^{\infty}\exp\left(-\frac{\pi^2 l^2}{x}\right)\cos 2\pi l\upsilon, \quad x > 0 \cdot$$

The sum by k is reduced to Bessel's modified function of the first kind [70]

$$I_0(x) = \frac{1}{\pi}\int_0^{\pi} d\phi \cdot e^{x\cdot\cos\phi}.$$

As a result, solution of Eq. (2.3.12) has the form

$$\mu = \frac{1}{\beta} \times$$

$$\times \ln\left\{\frac{Nd}{2L}\sqrt{\frac{\beta\varepsilon_0}{\pi}}e^{\beta\Delta}\left[ch\beta\mu_B B \cdot I_0(\beta\Delta)\left(1+2\sum_{l=1}^{\infty}\exp\left(-\frac{\pi^2 l^2}{\beta\varepsilon_0}\right)\cos 2\pi l\frac{\Phi}{\Phi_0}\right)\right]^{-1}\right\}.$$

(2.3.13)

This shows that chemical potential undergoes Aharonov-Bohm type oscillations with a change in magnetic field crossing the tube. De Haas-van Alphen type oscillations are not present in this case. In the absence of a superlattice they were considered in article [60].

Energy of an electron gas can be calculated by equation

$$E = -N\frac{\partial}{\partial\beta}\ln\sum_{mk\sigma}\exp\left(-\beta\varepsilon_{mk\sigma}\right).$$

It equals

$$E = \frac{N}{2\beta}\left\{2\beta\varepsilon_0\left\langle\left(m+\frac{\Phi}{\Phi_0}\right)^2\right\rangle+1-\right.$$

$$\left.-2\beta\Delta\left[1-\frac{I_0'(\beta\Delta)}{I_0(\beta\Delta)}\right]-2\beta\mu_B B th\beta\mu_B B\right\},$$

(2.3.14)

where

$$\langle P_m\rangle = \frac{\sum_m P_m\exp\left(-\beta\varepsilon_m\right)}{\sum_m\exp\left(-\beta\varepsilon_m\right)}, \quad \varepsilon_m = \varepsilon_0\left(m+\frac{\Phi}{\Phi_0}\right)^2.$$

Derivative with respect to the argument of the Bessel function is marked with a prime (').

Heat capacity of an electron gas equals

Electron Gas on the Surface of a Nanotube

$$C = \frac{N}{2}\left\{ 2(\beta\varepsilon_0)^2 \left[\left\langle \left(m + \frac{\Phi}{\Phi_0}\right)^4 \right\rangle - \left\langle \left(m + \frac{\Phi}{\Phi_0}\right)^2 \right\rangle^2 \right] + 1 + \right.$$
$$+ 2(\beta\Delta)^2 \left[I_0''(\beta\Delta)I_0(\beta\Delta) - \left(I_0'(\beta\Delta)\right)^2 \right] \times$$
$$\left. \times \left[\left[I_0(\beta\Delta)\right]^{-2} + 2\left(\frac{\beta\mu_B B}{ch\beta\mu_B B} \right)^2 \right] \right\}. \tag{2.3.15}$$

Separate terms in Eqs. (2.3.14) and (2.3.15) agree with the energy term in (2.3.1). The first term on the right side of expression (2.3.14) represents the average energy of centripetal motion of electrons at the surface of the nanotube, the second and third terms are due to longitudinal motion of electrons along the tube, and the last term is due to spin splitting of energy levels of an electron in a magnetic field. It coincides with the energy of a two-level system with distance $2\mu_B B$ between the levels. Expression (2.3.15) shows that the presence of a magnetic field does not affect the heat capacity term present due to electron motion. At the same time, modulation does not affect heat capacity related to centripetal motion of electrons and spin splitting of levels.

Using the presentation of a Bessel function as a row and its asymptote, "longitudinal" component of heat capacity (2.3.15) is confirmed to be equal to

$$C_{\parallel} = \frac{N}{2}\left\{ 1 + 2(\beta\Delta)^2 \left[I_0''(\beta\Delta)I_0(\beta\Delta) - \left(I_0'(\beta\Delta)\right)^2 \right] \cdot \left[I_0(\beta\Delta)\right]^{-2} \right\} =$$
$$= \begin{cases} \dfrac{N}{2}, \beta\Delta \ll 1, \\ N, \beta\Delta \gg 1. \end{cases}$$

This result agrees with the classical theory on equipartition of energy about degrees of freedom [71]. Its physical meaning is obvious. If energy of thermal motion of electrons β^{-1} is small compared to the modulating potential amplitude, the electrons oscillate slightly in the modulating potential gaps. These oscillations make a contribution to heat capacity in the amount of N. If β^{-1} exceeds modulation amplitude Δ, the electrons move freely along the tube. Contribution of this motion to heat capacity is equal to $N/2$. Thus, term $C_{\parallel}$ changes from N to $N/2$ as temperature increases. "Transverse" part of heat capacity depends on the magnetic flux. In weak magnetic fields, the inequality holds $\Phi \ll \Phi_0$. This allows the dismissal of magnetic field influence on the "transverse" component of heat capacity $C_{\perp}$. Then the following limiting expressions can be obtained:

$$C_{\perp} = \begin{cases} N/2, \beta\varepsilon_0 \ll 1, \\ N(\beta\varepsilon_0)^2 \exp(-\beta\varepsilon_0), \beta\varepsilon_0 \gg 1. \end{cases}$$

As expected, high temperature limit of $C_{\perp}$ is in accord with the theorem on equipartition of energy about degrees of freedom.

Conclusion

The theoretical calculation of density of electron states, chemical potential, internal energy, heat capacity, entropy of degenerated and nondegenerated electron gas on the surface of a cylindrical nanotube using the approximation of effective mass are presented in Subsection 2.1. We proved that thermodynamic functions of degenerated gas oscillate similarly to the de Haas-van Alphen oscillations accompanied with variation of chemical potential of electrons. An effective mass of electron and a radius of the nanotube are determines the period of this oscillations. Also we predicted the maximum of temperature dependence of nondegenerated electron gas heat capacity. The limiting transition to two-dimensional electron gas is carried out.

Electron Gas on the Surface of a Nanotube 43

Thermodynamic functions have been calculated in the Subsection 2.2 within the framework of the effective mass approximation for degenerate and nondegenerate electron gases on the semiconductor cylindrical nanotube surface in a longitudinal magnetic field. The Laplace transform linking the density of states and the statistical sum has been used. The thermodynamic quantities of degenerate electron gas undergo the de Haas-van Alphen oscillations with the electron Fermi energy change and the Aharonov-Bohm oscillations with the magnetic flux change within the semiconductor tube cross section. The quantities related to nondegenerate gas oscillate only with the change of magnetic flux. A peak has been found in the nondegenerate gas heat capacity-temperature diagram. A limiting process to 2D electron gas on plane has been carried out.

The effect of modulating potential at the surface of a nanotube in a longitudinal magnetic field on heat capacity of a degenerate and nondegenerate electron gas is considered in the Subsection 2.3. The heat capacity is represented by monotonic and oscillating terms. Heat capacity of a degenerate electron gas exhibits de Haas-van Alphen type oscillations, dependent on density of electrons and Aharonov-Bohm type oscillation dependent on the intensity of the magnetic field going through the nanotube cross section.

3. DYNAMICAL CONDUCTIVITY OF ELECTRON GAS ON THE TUBE

3.1. Dynamic Conductivity Tensor of Electron Gas on the Nanotube Surface in a Magnetic Field

Confined by the nanotube's cylindrical surface electron gas imbedded into a parallel to the cylinder axis magnetic field, has axial symmetry. Therefore, it is suitable to characterize the electron state by the conserved quantities: angular momentum projection $m = 0, \pm 1, \ldots$ and impulse projection k of electron on tube axis z, spin quantum number $\sigma = \pm 1$.

Electron stationary state $|mk\sigma\rangle$ wave function and its energy spectrum are equal [37 – 40] to

$$\Psi_{mk\sigma}(\varphi,z,\alpha)=\frac{1}{\sqrt{S}}e^{i(m\varphi+kz)}\chi_\sigma(\alpha),\qquad(3.1.1)$$

$$\varepsilon_{mk\sigma}=\varepsilon_0\left(m+\frac{\Phi}{\Phi_0}\right)^2+\frac{k^2}{2m_*}+\sigma\mu_B B,\qquad(3.1.2)$$

where m_* and μ_B are the effective mass and the electron spin magnetic moment, φ and z are the cylindrical coordinates, $\chi_\sigma(\alpha)$ is a spin wave function, $\varepsilon_0=1/2m_*a^2$ is a rotational quantum, a is the tube radius, $\Phi=\pi a^2 B$ is the magnetic induction B flux through the tube section, $\Phi_0=2\pi c/e$ is flux quantum, $S=2\pi aL$ is the nanotube lateral surface area, the length of nanotube is L. Hereinafter, the quantum constant is considered to equal unity. The spectrum (3.1.2) is a set of one-dimensional subzones whose boundaries $\varepsilon_{m0\sigma}$ are not equidistant.

One can expand the current electron density operator in field B into series of cylindrical harmonics (3.1.1):

$$\mathbf{J}(\varphi,z)=\sum_{mk}\mathbf{J}(m,k)\Psi_{mk}(\varphi,z),$$

where

$$\mathbf{J}(m,k)=e\sum_{m_1k_1m_2k_2\sigma}\langle m_1k_1\,|\,\mathbf{V}(-m,-k)\,|\,m_2k_2\rangle a^+_{m_1k_1\sigma}a_{m_2k_2\sigma}.\qquad(3.1.3)$$

Here $a_{mk\sigma}$ and $a^+_{mk\sigma}$ are annihilation and creation operators of electrons in quantum state $|mk\sigma\rangle$, $\langle m_1 k_1 | \mathbf{V}(-m,-k) | m_2 k_2 \rangle$ are operator $\mathbf{V}$ matrix elements

$$\mathbf{V}(-m,-k) = \frac{1}{2m_*}\left[\left(-i\nabla - \frac{e}{c}\mathbf{A}\right)\Psi^*_{mk}(\varphi,z) + \Psi^*_{mk}(\varphi,z)\left(-i\nabla - \frac{e}{c}\mathbf{A}\right)\right]$$

in the functional basis (3.1.1), $\mathbf{B} = rot\mathbf{A}$. The operator's (3.1.3) cylindrical components are equal to

$$J_\varphi(m,k) = \frac{e}{m_* a\sqrt{S}}\sum_{m'k'\sigma}\left(m' + \frac{\Phi}{\Phi_0} + \frac{m}{2}\right)a^+_{m'k'\sigma}a_{(m'+m)(k'+k)\sigma},$$

$$J_z(m,k) = \frac{e}{m_*\sqrt{S}}\sum_{m'k'\sigma}\left(k' + \frac{k}{2}\right)a^+_{m'k'\sigma}a_{(m'+m)(k'+k)\sigma}.$$

(3.1.4)

Within the framework of electron gas linear response theory [83, 84] to the weak electric field $\mathbf{E}$ varying with the wave vector q and frequency ω one can obtain the current density:

$$j_\mu(m,q,\omega) = \sum_\nu \sigma_{\mu\nu}(m,q,\omega)E_\nu(m,q,\omega),$$

where

$$\sigma_{\mu\nu}(m,q,\omega) = i\frac{e^2 n}{m_*\omega}\delta_{\mu\nu} +$$

$$+\frac{1}{\omega}\int_0^\infty dt\, e^{i\omega t}\left\langle\left[J_\mu(m,q,t), J_\nu(-m,-q,0)\right]\right\rangle$$

(3.1.5)

is the Kubo formula for conductivity tensor. Here $J_\mu(m,q,t)$ is a component of Heisenberg operator (3.1.4), n is a surface electron density,

$[a,b] = ab - ba$, angle brackets mark Gibbs average. We have substituted (3.1.4) in (3.1.5) and obtained the relation between the conductivity tensor and Fourier time component of two-electron Greens function [83, 84]:

$$K_t(1,2;3,4) = -i\theta(t)\left\langle\left[a_1^+(t)a_2(t), a_3^+(0)a_4(0)\right]\right\rangle, \quad (3.1.6)$$

where $1 = (m_1, k_1, \sigma_1), \ldots$, $\theta(t)$ is a Heaviside's function. In case of free electron gas, the function (3.1.6) is expressed by one-electron Green function. In the result, the conductivity tensor components (3.1.5) are equal

$$\sigma_{\varphi\varphi}(m,q,\omega) = i\frac{e^2 n}{m_*\omega} + \frac{ie^2}{m_*^2 a^2 \omega S} \cdot$$

$$\cdot \sum_{m'k\sigma} f(\varepsilon_{m'k\sigma}) \left[\frac{\left(m' + \frac{\Phi}{\Phi_0} + \frac{m}{2}\right)^2}{\varepsilon_{m'k\sigma} - \varepsilon_{(m'+m)(k+q)\sigma} + \omega + i0} - \frac{\left(m' + \frac{\Phi}{\Phi_0} - \frac{m}{2}\right)^2}{\varepsilon_{(m'-m)(k-q)\sigma} - \varepsilon_{m'k\sigma} + \omega + i0} \right],$$

$$\sigma_{\varphi z}(m,q,\omega) = \sigma_{z\varphi}(m,q,\omega) = \frac{ie^2}{m_*^2 a\omega S} \sum_{m'k\sigma} f(\varepsilon_{m'k\sigma}) \cdot$$

$$\cdot \left[\frac{\left(m' + \frac{\Phi}{\Phi_0} + \frac{m}{2}\right)\left(k + \frac{q}{2}\right)}{\varepsilon_{m'k\sigma} - \varepsilon_{(m'+m)(k+q)\sigma} + \omega + i0} - \frac{\left(m' + \frac{\Phi}{\Phi_0} - \frac{m}{2}\right)\left(k - \frac{q}{2}\right)}{\varepsilon_{(m'-m)(k-q)\sigma} - \varepsilon_{m'k\sigma} + \omega + i0} \right], \quad (3.1.7)$$

$$\sigma_{zz}(m,q,\omega) = i\frac{e^2 n}{m_*\omega} + \frac{ie^2}{m_*^2 \omega S} \sum_{m'k\sigma} f(\varepsilon_{m'k\sigma}) \cdot$$

$$\cdot \left[\frac{\left(k + \frac{q}{2}\right)^2}{\varepsilon_{m'k\sigma} - \varepsilon_{(m'+m)(k+q)\sigma} + \omega + i0} - \frac{\left(k - \frac{q}{2}\right)^2}{\varepsilon_{(m'-m)(k-q)\sigma} - \varepsilon_{m'k\sigma} + \omega + i0} \right],$$

Electron Gas on the Surface of a Nanotube

where $f(\varepsilon)$ is Fermi function. These components are included into the dispersion equation for electromagnetic waves propagating along the tube.

In case of $a \to \infty$ one can obtain from Eq. (3.1.7) the well-known Drude-Lorentz expression which describes two-dimensional electron gas with the magnetic field absence: $\sigma_{\varphi\varphi} = \sigma_{zz} = ie^2 n / m_* \omega$, $\sigma_{\varphi z} = 0$. The real part of conductivity tensor in this case is equal to zero. It is caused by not taking into account the electrons collisions. That effect can be considered by means of $\omega \to \omega + i\nu_c$ replacement in Eq. (3.1.7). Here ν_c is the collisions frequency.

At zero temperature, the integrals over k included into Eq. (3.1.7) can be calculated exactly. Thus, real and imaginary parts of conductivity (3.1.7) are equal to

$$\operatorname{Re}\sigma_{\varphi\varphi}(m,q,\omega) = -\frac{e^2}{4\pi m_* |q| a^3 \omega} \times$$

$$\times \sum_{m'\sigma} \left\{ \left(m' + \frac{\Phi}{\Phi_0} + \frac{m}{2} \right)^2 \left[\theta(A_{+-+}) - \theta(A_{--+}) \right] - \left(m' + \frac{\Phi}{\Phi_0} - \frac{m}{2} \right)^2 \left[\theta(A_{++-}) - \theta(A_{-+-}) \right] \right\},$$

$$\operatorname{Re}\sigma_{\varphi z}(m,q,\omega) = -\frac{e^2}{4\pi (qa)^2 \omega} \times$$

$$\times \sum_{m'\sigma} \left\{ \left(m' + \frac{\Phi}{\Phi_0} + \frac{m}{2} \right)(\omega - \Omega_+) \left[\theta(A_{+-+}) - \theta(A_{--+}) \right] - \left(m' + \frac{\Phi}{\Phi_0} - \frac{m}{2} \right)(\omega - \Omega_-) \left[\theta(A_{++-}) - \theta(A_{-+-}) \right] \right\},$$

$$\operatorname{Re}\sigma_{zz}(m,q,\omega) = -\frac{m_* e^2}{4\pi |q|^3 a\omega} \times$$

$$\times \sum_{m'\sigma} \left\{ (\omega - \Omega_+)^2 \left[\theta(A_{+-+}) - \theta(A_{--+}) \right] - (\omega - \Omega_-)^2 \left[\theta(A_{++-}) - \theta(A_{-+-}) \right] \right\},$$

(3.1.8)

$$\operatorname{Im}\sigma_{\varphi\varphi}(m,q,\omega)=\frac{e^2n}{m_*\omega}-\frac{e^2}{4\pi^2 m_*qa^3\omega}\times$$

$$\times\sum_{m'\sigma}\left[\left(m'+\frac{\Phi}{\Phi_0}+\frac{m}{2}\right)^2\ln\left|\frac{A_{-++}}{A_{-+}}\right|-\left(m'+\frac{\Phi}{\Phi_0}-\frac{m}{2}\right)^2\ln\left|\frac{A_{++-}}{A_{+-}}\right|\right],$$

$$\operatorname{Im}\sigma_{\varphi z}(m,q,\omega)=-\frac{e^2mn}{m_*qa\omega}-\frac{e^2}{4\pi^2(qa)^2\omega}\times$$

$$\times\sum_{m'\sigma}\left[\left(m'+\frac{\Phi}{\Phi_0}+\frac{m}{2}\right)(\omega-\Omega_+)\ln\left|\frac{A_{-++}}{A_{-+}}\right|-\left(m'+\frac{\Phi}{\Phi_0}-\frac{m}{2}\right)(\omega-\Omega_-)\ln\left|\frac{A_{++-}}{A_{+-}}\right|\right],$$

$$\operatorname{Im}\sigma_{zz}(m,q,\omega)=-\frac{2e^2\varepsilon_0 m^2n}{q^2\omega}-\frac{m_*e^2}{4\pi^2q^3a\omega}\times$$

$$\times\sum_{m'\sigma}\left[(\omega-\Omega_+)^2\ln\left|\frac{A_{-++}}{A_{-+}}\right|-(\omega-\Omega_-)^2\ln\left|\frac{A_{++-}}{A_{+-}}\right|\right],$$

(3.1.9)

where

$$A_{+-+}=q\upsilon_{m'}^\sigma-\omega_-+\Omega_+,\quad A_{--+}=-q\upsilon_{m'}^\sigma-\omega_-+\Omega_+,$$

$$A_{++-}=q\upsilon_{m'}^\sigma-\omega_++\Omega_-,\quad A_{+--}=-q\upsilon_{m'}^\sigma-\omega_++\Omega_-,$$

$$\upsilon_{m'}^\sigma=\sqrt{\frac{2}{m_*}}\left(\varepsilon_F-\varepsilon_{m'\sigma}\right)^{1/2}$$

is the electron motion along the tube axis velocity maximum in subzone (m',σ), $\varepsilon_{m'\sigma}=\varepsilon_{m'0\sigma}$ is the subzone boundary, ε_F is the Fermi energy, $\omega_\pm=\omega\pm q^2/2m_*$,

$$\Omega_\pm(m,m')=\varepsilon_0\left[\pm\left(m'+\frac{\Phi}{\Phi_0}\pm m\right)^2\mp\left(m'+\frac{\Phi}{\Phi_0}\right)^2\right]$$

is the frequency of the «vertical» electron transitions between the spectrum (3.1.2) subzones. While obtaining (3.1.9) out of (3.1.7) the identity below was used

$$\frac{B^2}{B-C} = B + C + \frac{C^2}{B-C}.$$

Here we concentrates on the analysis of degenerate electron gas longitudinal conductivity (3.1.8), (3.1.9) assuming $m = 0$. The numerical calculations were carried out also for $m = 1$ case. In this case electrons make only intra-subzone transitions without spin change. Longitudinal conductivity is included into the propagating along the nanotube axis intra-subzone magnetoplasma waves dispersion equation [39] because it is related to the polarization operator P by the relation below

$$\operatorname{Im}\sigma_{zz}(0,q,\omega) = \frac{e^2 \omega}{q^2} \operatorname{Re} P(0,q,\omega).$$

In case of $m = 0$ we obtain the following equations from (3.1.8) and (3.1.9)

$$\operatorname{Re}\sigma_{zz}(q,\omega) = -\frac{e^2 m_* \omega}{4\pi |q|^3 a} \times$$

$$\times \sum_{m'\sigma}\left[\theta\left(q\upsilon_{m'}^\sigma - \omega_-\right) - \theta\left(-q\upsilon_{m'}^\sigma - \omega_-\right) - \theta\left(q\upsilon_{m'}^\sigma - \omega_+\right) + \right. \qquad (3.1.10)$$

$$\left. + \theta\left(-q\upsilon_{m'}^\sigma - \omega_+\right)\right],$$

$$\operatorname{Im}\sigma_{zz}(q,\omega) = -\frac{e^2 m_* \omega}{4\pi^2 q^3 a} \sum_{m'\sigma}\left(\ln\left|\frac{q\upsilon_{m'}^\sigma - \omega_-}{-q\upsilon_{m'}^\sigma - \omega_-}\right| - \ln\left|\frac{q\upsilon_{m'}^\sigma - \omega_+}{-q\upsilon_{m'}^\sigma - \omega_+}\right|\right).$$

Suppose that the flux ratio in (3.1.2) should be equal to $\Phi/\Phi_0 = M + \eta$, where $M = 0, 1, \ldots$ is an integer part of Φ/Φ_0, η is a fractional part ($0 \leq \eta < 1$). One can use the achievable in experiments values of carbon and semiconductor nanotube radii and obtain that the value of M in the accessible magnetic fields is low, unless the tube radius is too large. Then the energy of electron (3.1.2) is minimal in the subzone $(m' = -M, \sigma = -1)$:

$$\varepsilon_{min} = \varepsilon_0 \eta^2 - \mu_B B.$$

In this case the electron rotation energy is nearly compensated by the rotation in the magnetic field. Usually spin level splitting (3.1.2) is small, i.e., $2\mu_B B < \varepsilon_0 (1 + 2\eta)$

and the electron density corresponds to inequality

$$n < \frac{\sqrt{m_* \mu_B B}}{\pi^2 a}.$$

Then electrons partially fill only the lower subzone $(-M, -1)$. Therefore, equation (3.1.10) leads to the following

$$\mathrm{Re}\, \delta\sigma = -\frac{\pi}{2} \frac{k_0}{q} x^2, \tag{3.1.11}$$

$$\mathrm{Im}\, \delta\sigma = -\frac{k_0 x^2}{2q} \ln \left| \frac{(1 + q/2k_0)^2 - x^2}{(1 - q/2k_0)^2 - x^2} \right|, \tag{3.1.12}$$

where

$$\delta\sigma = \sigma_{zz}(q,\omega) \frac{m_* \omega}{e^2 n}, \quad x = \frac{\omega}{q \upsilon_0}, \quad \upsilon_0 = \frac{k_0}{m_*} = \sqrt{\frac{2}{m_*}} \left(\varepsilon_F + \mu_B B - \varepsilon_0 \eta^2 \right)^{1/2}$$

is the electrons velocity maximum in subzone $(-M,-1)$. Figures 6, 7 shows the $x = \omega/qv_0$ dependencies of $\text{Re}\,\delta\sigma$ and $\text{Im}\,\delta\sigma$ under $\eta = 0.01, q = k_0, k_0 a = 1$ in quantum limit when Fermi energy is located in subzone $(-M,-1)$.

One can calculate the included into Eq. (3.1.10) sum by m' using a Poisson formula assuming $\varepsilon_F \gg \varepsilon_0$, i.e., a large quantity of subzones are filled. Within long wavelength assumption $qv_F \ll \omega$ (v_F is a Fermi velocity) we obtain the following formula from Eq. (3.1.10)

$$\text{Im}\,\sigma_{zz} = \frac{e^2 n}{m_* \omega} \times$$
$$\times \left[1 + \frac{2}{\pi^2}\left(\frac{\varepsilon_0}{\varepsilon_F}\right)^{3/4} \sum_{l=1}^{\infty} \frac{1}{l^{3/2}} \sin\left(2\pi l \sqrt{\frac{\varepsilon_F}{\varepsilon_0}} - \frac{\pi}{4}\right) \cos\left(2\pi l \frac{\Phi}{\Phi_0}\right)\right]. \quad (3.1.13)$$

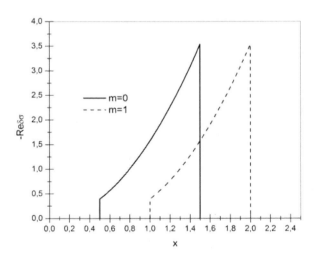

Figure 6. Dependence of $\text{Re}\,\delta\sigma$ on $x = \omega/qv_0$ under $\eta = 0.01, q = k_0, k_0 a = 1$ and $m = 0$ (solid curve), $m = 1$ (dashed).

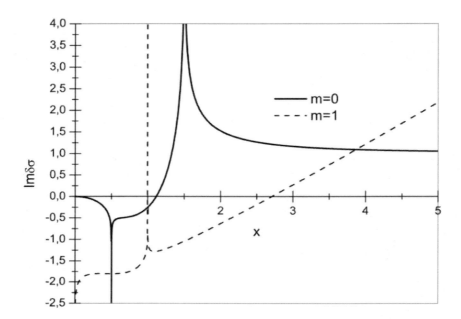

Figure 7. Dependence of $\operatorname{Im}\delta\sigma$ on $x = \omega/q\upsilon_0$ under $\eta = 0.01, q = k_0, k_0 a = 1$ and $m = 0$ (solid curve), $m = 1$ (dashed).

Here we neglected the spin level splitting (3.1.2). Monotonous part of conductivity (3.1.13) under the given electron density does not depend on the tube radius and coincides with the two-dimensional electron gas conductivity [73]. It is within the $a \to \infty$ when the tube is cut along its generatrix and unrolled onto a plane. Conductivity (3.1.13) experiences the Aharonov-Bohm and de Haas - van Alphen oscillation types. The former is related to the pass of electron states density root peculiarities on the subzone boundaries throw Fermi energy with the tube radius or electron density changes. If we study conductivity dependence (3.1.13) on $\varepsilon_F^{1/2} = (n/\nu)^{1/2}$ (ν is the two-dimensional electron gas states density), then the oscillation period is $1/\sqrt{2m_* a}$.

Electron Gas on the Surface of a Nanotube 53

If we study the radius dependence, then the period of conductivity oscillations is equal to $1/\sqrt{2m_*\varepsilon_F}$. Aharonov-Bohm oscillations are caused by the magnetic flux through the tube cross-section change. Their period is equal to the flux quantum. Figure 8 shows the dependence of the Eq. (3.1.13) first conductivity harmonic on $(\varepsilon_F/\varepsilon_0)^{1/2}$ under $\Phi/\Phi_0 = 0,01$.

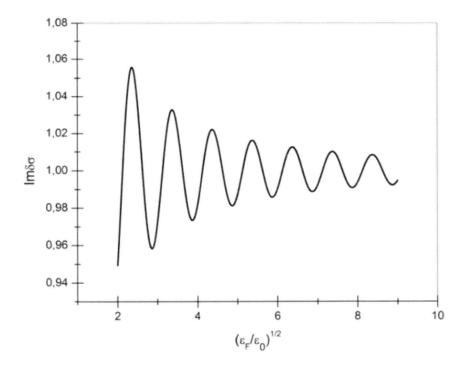

Figure 8. Dependence of the first harmonic $\mathrm{Im}\,\delta\sigma$ (3.1.13) on $(\varepsilon_F/\varepsilon_0)^{1/2}$ under $\Phi/\Phi_0 = 0,01$.

Nondegenerate electron gas emerges if inequality $\beta|\mu| \gg 1$ is true on the semiconductor nanotube surface. Here β is a reverse temperature, μ is a chemical potential. Using Boltzmann's statistics, the chemical potential is related to the electron density as

$$n = \frac{1}{\pi^{3/2} a} \sqrt{\frac{m_*}{2\beta}} e^{\beta\mu} \operatorname{ch}\left(\beta\mu_B B\right) \sum_{m=-\infty}^{\infty} e^{-\beta\varepsilon_m}, \qquad (3.1.14)$$

where $\varepsilon_m = \varepsilon_0 \left(m + \Phi/\Phi_0\right)^2$. The included here summation may be transformed by means of formula Ref. [69]

$$\sum_{m=-\infty}^{\infty} e^{-x(m+\upsilon)^2} = \sqrt{\frac{\pi}{x}} \sum_{l=-\infty}^{\infty} e^{-\frac{\pi^2 l^2}{x}} \cos\left(2\pi l\upsilon\right), \quad x > 0. \qquad (3.1.15)$$

Real and imaginary tensor (3.1.7) parts in the studied case assuming $m = 0$ are equal to

$$\operatorname{Re}\sigma_{\varphi\varphi}\left(q,\omega\right) = \frac{e^2}{4\sqrt{\pi} m_* \left(\beta\varepsilon_0\right)^{3/2} |q| a^3 \omega} e^{\beta\mu} \times$$

$$\times \operatorname{ch}\left(\beta\mu_B B\right)\left\{\exp\left[-\frac{m_*\beta}{2}\left(\frac{\omega_-}{q}\right)^2\right] - \exp\left[-\frac{m_*\beta}{2}\left(\frac{\omega_+}{q}\right)^2\right]\right\} \times$$

$$\times \sum_{l=-\infty}^{\infty} \exp\left(-\frac{\pi^2 l^2}{\beta\varepsilon_0}\right) \times \left(1 - \frac{2\pi^2 l^2}{\beta\varepsilon_0}\right)\cos\left(2\pi l \frac{\Phi}{\Phi_0}\right),$$

$$\operatorname{Re}\sigma_{\varphi z}\left(q,\omega\right) = \frac{\sqrt{\pi} e^2}{2\left(\beta\varepsilon_0\right)^{3/2} \left(qa\right)^2} e^{\beta\mu} \operatorname{ch}\left(\beta\mu_B B\right) \times$$

$$\times \sum_{l=-\infty}^{\infty} l \exp\left(-\frac{\pi^2 l^2}{\beta\varepsilon_0}\right)\left\{\exp\left[-\frac{m_*\beta}{2}\left(\frac{\omega_-}{q}\right)^2\right] -\right.$$

$$\left. - \exp\left[-\frac{m_*\beta}{2}\left(\frac{\omega_+}{q}\right)^2\right]\right\}\sin\left(2\pi l \frac{\Phi}{\Phi_0}\right),$$

$$\operatorname{Re}\sigma_{zz}(q,\omega)=\frac{e^2 n\omega}{|q|^3}\sqrt{\frac{\pi m_*\beta}{2}}\left\{\exp\left[-\frac{m_*\beta}{2}\left(\frac{\omega_-}{q}\right)^2\right]-\exp\left[-\frac{m_*\beta}{2}\left(\frac{\omega_+}{q}\right)^2\right]\right\},$$

$$\operatorname{Im}\sigma_{\varphi\varphi}(q,\omega)=\frac{e^2 n}{m_*\omega}+\frac{e^2}{4\pi m_*(\beta\varepsilon_0)^{3/2}qa^3\omega}e^{\beta\mu}\operatorname{ch}(\beta\mu_B B)\left[F\left(\sqrt{\frac{m_*\beta}{2}}\frac{\omega_-}{q}\right)-\right.$$

$$\left.-F\left(\sqrt{\frac{m_*\beta}{2}}\frac{\omega_+}{q}\right)\right]\sum_{l=-\infty}^{\infty}\exp\left(-\frac{\pi^2 l^2}{\beta\varepsilon_0}\right)\left(1-\frac{2\pi^2 l^2}{\beta\varepsilon_0}\right)\cos\left(2\pi l\frac{\Phi}{\Phi_0}\right),$$

$$\operatorname{Im}\sigma_{\varphi z}(q,\omega)=\frac{e^2}{2(\beta\varepsilon_0)^{3/2}(aq)^2}e^{\beta\mu}\operatorname{ch}(\beta\mu_B B)\times$$

$$\times\sum_{l=-\infty}^{\infty}l\exp\left(-\frac{\pi^2 l^2}{\beta\varepsilon_0}\right)\left\{F\left(\sqrt{\frac{m_*\beta}{2}}\frac{\omega_-}{q}\right)-F\left(\sqrt{\frac{m_*\beta}{2}}\frac{\omega_+}{q}\right)\right\}\sin\left(2\pi l\frac{\Phi}{\Phi_0}\right),$$

$$\operatorname{Im}\sigma_{zz}(q,\omega)=\frac{e^2 n\omega}{q^3}\sqrt{\frac{m_*\beta}{2}}\left[F\left(\sqrt{\frac{m_*\beta}{2}}\frac{\omega_-}{q}\right)-F\left(\sqrt{\frac{m_*\beta}{2}}\frac{\omega_+}{q}\right)\right],$$

$$(3.1.16)$$

where

$$F(x)=\frac{1}{\sqrt{\pi}}\operatorname{P}\int_{-\infty}^{\infty}dy\,\frac{e^{-y^2}}{x-y}.$$

Components $\sigma_{\varphi\varphi}$ and $\sigma_{\varphi z}$ in Eq. (3.1.16) are expressed through β and μ, i.e., they refer to an open system of tube surface electrons. They experience only Aharonov-Bohm oscillations with the magnetic field changes. Aharonov-Bohm oscillations are not taking place in σ_{zz} component while it is described in β and n terms.

3.2. Electron Gas High-Frequency Conductivity on the Surface of a Nanotube with Superlattice in Magnetic Field

For the nanotube with superlattice in magnetic field, the surface electron gas linear response to an electromagnetic wave

$$\vec{E} = \vec{E}_0 \exp \mathrm{i}\left(m\varphi + qz - \omega t\right)$$

is characterized by conductivity two-dimensional tensor $\sigma_{\alpha\beta}\left(m,q,\omega\right)$. Here $\vec{E}$ is the electric field of wave, m is the integer number, q and ω are the wave vector and frequency of the wave, φ and z are cylindrical coordinates. The density of surface current on the tube is

$$j_\alpha\left(m,q,\omega\right) = \sum_\beta \sigma_{\alpha\beta}\left(m,q,\omega\right) E_\beta\left(m,q,\omega\right), \tag{3.2.1}$$

where $j_\alpha\left(m,q,\omega\right)$ and $E_\beta\left(m,q,\omega\right)$ are cylindrical harmonics of $\vec{j}$ and $\vec{E}$ vectors. Kubo's formula for the conductivity tensor of electron gas on the surface of the nanotube with superlattice is [62]

$$\sigma_{\alpha\beta}\left(m,q,\omega\right) = \mathrm{i}\frac{e^2 n}{m_* \omega}\delta_{\alpha\beta} + \frac{1}{\omega}\int_0^\infty dt\, e^{\mathrm{i}\omega t}\left\langle\left[J_\alpha\left(m,q,t\right), J_\beta\left(-m,-q,0\right)\right]\right\rangle, \tag{3.2.2}$$

where m_* and e are, respectively, electron effective mass and charge, n is surface density of electrons, and $\vec{J}\left(m,q,t\right)$ is the cylindrical harmonics of current density operator in the external magnetic field $\vec{B}$. The angle brackets denote the average value of the operator commutator. The quantum constant was assumed as unity. The components of $\vec{J}\left(m,q\right)$ vector are

$$\begin{aligned} J_\varphi &= -\frac{2e}{m_* a\sqrt{S}}\sum_{lk}\left(l+\eta+\frac{m}{2}\right)a_{lk}^+ a_{(l+m)(k+q)}, \\ J_z &= -\frac{2e}{m_*\sqrt{S}}\sum_{lk}\left(k+\frac{q}{2}\right)a_{lk}^+ a_{(l+m)(k+q)}, \end{aligned} \tag{3.2.3}$$

Electron Gas on the Surface of a Nanotube 57

where l and k are projections of electron angular moment and momentum, respectively, onto the axis of the tube with radius a, a_{lk} and a_{lk}^+ are operators of annihilation and creation of electrons in $|lk\rangle$ state, $\eta = \Phi/\Phi_0$ is the ratio of magnetic flux $\Phi = \pi a^2 B$ through the tube cross-section to the flux quantum $\Phi_0 = 2\pi c/e$ [74], $S = 2\pi aL$ is the surface area for the tube with length L. Spin splitting of levels is not considered in Eq. (3.2.3).

From expressions (3.2.2) and (3.2.3) we obtain the components of conductivity tensor:

$$\sigma_{\varphi\varphi} = i\frac{e^2 n}{m_* \omega} + i\frac{2e^2}{m_*^2 a^2 \omega S} \times$$

$$\times \sum_{lk} f(\varepsilon_{lk}) \left[\frac{\left(l + \eta + \frac{m}{2}\right)^2}{\varepsilon_{lk} - \varepsilon_{(l+m)(k+q)} + \omega + i0} - \frac{\left(l + \eta - \frac{m}{2}\right)^2}{\varepsilon_{(l-m)(k-q)} - \varepsilon_{lk} + \omega + i0} \right], \quad (3.2.4)$$

$$\sigma_{\varphi z} = \sigma_{z\varphi} = i\frac{2e^2}{m_*^2 a\omega S} \times$$

$$\times \sum_{lk} f(\varepsilon_{lk}) \left[\frac{\left(l + \eta + \frac{m}{2}\right)\left(k + \frac{q}{2}\right)}{\varepsilon_{lk} - \varepsilon_{(l+m)(k+q)} + \omega + i0} - \frac{\left(l + \eta - \frac{m}{2}\right)\left(k - \frac{q}{2}\right)}{\varepsilon_{(l-m)(k-q)} - \varepsilon_{lk} + \omega + i0} \right], \quad (3.2.5)$$

$$\sigma_{zz} = i\frac{e^2 n}{m_* \omega} + i\frac{2e^2}{m_*^2 \omega S} \times$$

$$\times \sum_{lk} f(\varepsilon_{lk}) \left[\frac{\left(k + \frac{q}{2}\right)^2}{\varepsilon_{lk} - \varepsilon_{(l+m)(k+q)} + \omega + i0} - \frac{\left(k - \frac{q}{2}\right)^2}{\varepsilon_{(l-m)(k-q)} - \varepsilon_{lk} + \omega + i0} \right]. \quad (3.2.6)$$

Here f is Fermi function, ε_{lk} is electron energy on the tube surface. That is [31, 74, 85-87]

$$\varepsilon_{lk} = \varepsilon_0 (l+\eta)^2 + \Delta(1-\cos kd), \tag{3.2.7}$$

where $\varepsilon_0 = \left(2m_* a^2\right)^{-1}$ is rotational quantum, Δ and d are, respectively, amplitude and period of modulating potential on the tube surface. The first term in equation (3.2.7) was obtained in [74]. The second addend in the right part of (3.2.7) is taken from the theory of tight binding of electrons with the lattice. This is often used in the theory of semiconductor superlattices [31, 85, 86]. The real parts of the components $\sigma_{\varphi\varphi}$ and σ_{zz} are even functions of m and ω, while imaginary parts are odd ones.

At zero temperature in summation $\sum\limits_{k}$ the values k in the formulas (3.2.4)-(3.2.6) are limited to gap $-k_l \le k \le k_l$, where

$$k_l = \frac{1}{d} \arccos \frac{\varepsilon_l + \Delta - \mu}{\Delta}$$

is the maximum momentum of the electrons in the miniband l, $\varepsilon_l = \varepsilon_0 (l+\eta)^2$ is the miniband boundaries, μ is the Fermi energy. If $q = 0$, at zero temperature from the formulas (3.2.4)-(3.2.6) we calculate the components of dynamical conductivity tensor:

$$\mathrm{Re}\,\sigma_{\varphi\varphi}(m,\omega) = \frac{e^2}{\pi m_*^2 a^3 \omega} \times$$
$$\times \sum_l k_l \left[\left(l+\eta+\frac{m}{2}\right)^2 \delta(\omega-\Omega_+) - \left(l+\eta-\frac{m}{2}\right)^2 \delta(\omega-\Omega_-) \right], \tag{3.2.8}$$

$$\mathrm{Im}\,\sigma_{\varphi\varphi} = \frac{e^2 n}{m_*\omega} + \frac{e^2}{\pi^2 m_*^2 a^3 \omega} \sum_l k_l \left[\frac{\left(l+\eta+\frac{m}{2}\right)^2}{\omega-\Omega_+} - \frac{\left(l+\eta-\frac{m}{2}\right)^2}{\omega-\Omega_-} \right],$$

$$\sigma_{\varphi z}(m,\omega)=0,$$

$$\mathrm{Re}\,\sigma_{zz}(m,\omega)=\frac{e^2}{3\pi m_*^2 a\omega}\sum_l k_l^3\left[\delta(\omega-\Omega_+)-\delta(\omega-\Omega_-)\right],$$

$$\mathrm{Im}\,\sigma_{zz}(m,\omega)=\frac{e^2 n}{m_*\omega}+\frac{e^2}{3\pi^2 m_*^2 a\omega}\sum_l k_l^3\left[\frac{1}{\omega-\Omega_+}-\frac{1}{\omega-\Omega_-}\right].$$

(3.2.9)

Here

$$\Omega_{\pm}=\varepsilon_0 m\left[2(l+\eta)\pm m\right]$$

are frequencies of direct transitions of electrons between the miniband boundaries ε_l in the field of electromagnetic wave. During the transitions, conservation laws for longitudinal components of angular moment, momentum, and energy are satisfied.

At zero temperature, the summation over l in Eqs. (3.2.8) and (3.2.9) is limited by the condition $|\varepsilon_l+\Delta-\mu|\le\Delta$. This means that Fermi energy is concentrated within the miniband. The minibands are positioned in the intervals $[\varepsilon_l,\varepsilon_l+2\Delta]$ and have the width 2Δ.

Generally, the semiconductor nanotubes with radius $a\sim\left(10^{-7}-10^{-6}\right)cm$ in magnetic field $B\sim 10^5\,G$ are used. In this case, the electrons of the semiconductor nanotube occupy little quantity of bottom minibands, which boundaries at $\eta<1/2$ satisfy the inequality $\varepsilon_0\eta^2<\varepsilon_{-1}<\varepsilon_1<\varepsilon_{-2}<...$ In the quantum limit where $n<1/\pi ad$, Fermi energy is concentrated in the bottom miniband $l=0$ $\left[\varepsilon_0\eta^2,\varepsilon_0\eta^2+2\Delta\right]$. In this case, in the absence of spatial dispersion, from Eqs. (3.2.8) and (3.2.9) we obtain

$$\mathrm{Re}\,\sigma_{\varphi\varphi} = \frac{e^2 k_0}{\pi m_*^2 a^3 \omega} \times$$

$$\times \left[\left(\eta + \frac{m}{2} \right)^2 \delta\big(\omega - \varepsilon_0 m(2\eta + m)\big) - \left(\eta - \frac{m}{2} \right)^2 \delta\big(\omega - \varepsilon_0 m(2\eta - m)\big) \right], \quad (3.2.10)$$

$$\mathrm{Im}\,\sigma_{\varphi\varphi} = \frac{e^2 n}{m_* \omega} + \frac{e^2 k_0}{\pi^2 m_*^2 a^3 \omega} \left[\frac{\left(\eta + \dfrac{m}{2} \right)^2}{\omega - \varepsilon_0 m(2\eta + m)} - \frac{\left(\eta - \dfrac{m}{2} \right)^2}{\omega - \varepsilon_0 m(2\eta - m)} \right],$$

$$\mathrm{Re}\,\sigma_{zz} = \frac{e^2 k_0^3}{3\pi m_*^2 a \omega} \big[\delta\big(\omega - \varepsilon_0 m(2\eta + m)\big) - \delta\big(\omega - \varepsilon_0 m(2\eta - m)\big) \big],$$

$$\mathrm{Im}\,\sigma_{zz} = \frac{e^2 n}{m_* \omega} + \frac{e^2 k_0^3}{3\pi^2 m_*^2 a \omega} \left[\frac{1}{\omega - \varepsilon_0 m(2\eta + m)} - \frac{1}{\omega - \varepsilon_0 m(2\eta - m)} \right]. \quad (3.2.11)$$

Here $\Omega_\pm = \varepsilon_0 m(2\eta \pm m)$. The superlattice parameters Δ and d are included in Eqs. (3.2.10) and (3.2.11) only via the maximum momentum k_0 of electrons in the bottom miniband. In the absence of superlattice:

$$\Delta \to \infty, \ d \to 0, \ d^2 \Delta \to m_*^{-1}. \ \text{Then}$$

$$k_l = \left[2m_* (\mu - \varepsilon_l) \right]^{1/2},$$

and the Eqs. (3.2.10) and (3.2.11) agree with ones obtained in Ref. [62]. At $m = 0$, only the imaginary part $e^2 n / m_* \omega$ remains in Eqs. (3.2.10) and (3.2.11), while the real part is zero. This determines the electromagnetic wave energy absorbed by electrons. In the absence of direct and indirect transitions of electrons, the absorption is zero.

As the electron density grows, the number of addends in Eqs. (3.2.8) and (3.2.9) increases. If Fermi energy is concentrated in the second miniband, the oscillator forces of electron resonance transitions in Eqs.

Electron Gas on the Surface of a Nanotube   61

(3.2.8) and (3.2.9) are determined by values k_0 and k_{-1}. These are included in Eqs. (3.2.8) and (3.2.9) if the minibands are overlapped, i. e. $\varepsilon_0\eta^2 + 2\Delta > \varepsilon_{-1}$, and Fermi energy is concentrated in the overlap area $\left[\varepsilon_{-1}, \varepsilon_0\eta^2 + 2\Delta\right]$. Otherwise, the overlapping of minibands is absent. Then the maximum momentum of electrons k_0 in the completely occupied bottom miniband corresponds to Brillouin zone boundary π/d.

Taking into consideration weak spatial dispersion, the corrections $\mathrm{Im}\,\delta\sigma \sim q^2$ are induced into the imaginary parts of longitudinal and transversal conductivity. If $\Delta\sin qd/2 << |\omega - \Omega_{\pm}|$, these are the following

$$\mathrm{Im}\,\delta\sigma_{\varphi\varphi} = \frac{2e^2\Delta}{\pi^2 m_*^2 a^3 d\omega}\sin^2\frac{qd}{2}\sum_l \sin k_l d\left[\frac{\left(l+\eta+\dfrac{m}{2}\right)^2}{(\omega-\Omega_+)^2} + \frac{\left(l+\eta-\dfrac{m}{2}\right)^2}{(\omega-\Omega_-)^2}\right],$$

$$\mathrm{Im}\,\delta\sigma_{zz} = \frac{e^2 d\Delta q^2}{2\pi^2 m_*^2 a\omega}\sum_l k_l^2 \sin k_l d\left[\frac{1}{(\omega-\Omega_+)^2} + \frac{1}{(\omega-\Omega_-)^2}\right].$$

In the correction to the longitudinal conductivity also the condition $qd << 1$ was taken into account. These corrections are necessary to introduce into the dispersion equation for the spectrum of electromagnetic waves propagating along the tube.

In the quantum limit, taking into account the spatial dispersion, the real part of conductivity depends on Fermi level position in the bottom miniband. If μ is positioned in the bottom half of the miniband $\left(\varepsilon_0\eta^2 < \mu < \varepsilon_0\eta^2 + \Delta, q < \pi/2d\right)$ from Eq. (3.2.6) we obtain

$$\mathrm{Re}\,\sigma_{zz} = \frac{e^2\left(k^-\right)^2}{2\pi m_*^2 ad\omega}\left[4\Delta^2\sin^2\frac{qd}{2} - (\omega-\Omega_+)^2\right]^{-\frac{1}{2}},$$

where

$$k^- = \frac{1}{d}\arcsin\frac{|\omega-\Omega_+|}{2\Delta\sin\dfrac{qd}{2}}, \quad \omega_- < \omega < \omega_+,$$

$$\omega_\pm = \Omega_+ \pm 2\Delta\sin\frac{qd}{2}\left(\alpha_0\sin\frac{qd}{2} + \sqrt{1-\alpha_0^2}\cos\frac{qd}{2}\right),$$

$$\alpha_0 = \frac{\varepsilon_0\eta^2 + \Delta - \mu}{\Delta}.$$

The real parts of other components of conductivity tensor are obtained from $\mathrm{Re}\,\sigma_{zz}$ using substitution of $\left(k^-\right)^2$ by $(\eta+m/2)^2/a^2$ in $\mathrm{Re}\,\sigma_{\varphi\varphi}$ and by $k^-(\eta+m/2)/a$ in $\mathrm{Re}\,\sigma_{\varphi z}$.

If the Fermi level is positioned in the upper half of the miniband ($\varepsilon_0\eta^2 + \Delta < \mu < \varepsilon_0\eta^2 + 2\Delta$, $\pi/2d < q < \pi/d$), we obtain

$$\mathrm{Re}\,\sigma_{zz} = \frac{e^2\left(k^+\right)^2}{2\pi m_*^2 a d\omega}\left[4\Delta^2\sin^2\frac{qd}{2} - (\omega-\Omega_+)^2\right]^{-\frac{1}{2}},$$

where

$$k^+ = \frac{\pi}{d} - k^-, \quad \omega_- < \omega < \omega_+,$$

$$\omega_\pm = \Omega_+ \pm 2\Delta\sin\frac{qd}{2}\left(|\alpha_0|\sin\frac{qd}{2} + \sqrt{1-\alpha_0^2}\cos\frac{qd}{2}\right).$$

The real part of conductivity is nonzero in the area of Landau damping $\left[\omega_-,\omega_+\right]$ of electromagnetic waves in the tube.

Electron Gas on the Surface of a Nanotube 63

In the quasi-classical case, the quantization of electron circular rotation can be neglected. That is possible under condition of $\varepsilon_0 \ll \mu$. Substituting the l summation by integrals in Eqs. (3.2.8) and (3.2.9), we obtain

$$\text{Re}\,\sigma_{\varphi\varphi} = \frac{e^2\omega}{8\pi m_*^2 \left(a\varepsilon_0 |m|\right)^3}\left(k_+ I_+ - k_- I_-\right),$$

$$\text{Re}\,\sigma_{zz} = \frac{e^2}{6\pi m_*^2 a\varepsilon_0 |m|\omega}\left(k_+^3 I_+ - k_-^3 I_-\right),$$

(3.2.12)

where

$$k_\pm = \frac{1}{d}\arccos\frac{\varepsilon_\pm + \Delta - \mu}{\Delta},$$

$$\varepsilon_\pm = \varepsilon_0\left(\frac{\omega}{2m\varepsilon_0} \mp \frac{m}{2}\right)^2,$$

$$I_\pm(m,\omega) = \Theta\left(\frac{\omega}{2m\varepsilon_0} \mp \frac{m}{2} + \sqrt{\frac{\mu}{\varepsilon_0}}\right)\Theta\left(\sqrt{\frac{\mu}{\varepsilon_0}} - \frac{\omega}{2m\varepsilon_0} \pm \frac{m}{2}\right),$$

and Θ is Heaviside function.

If $\omega > 0$ and $|m| < 2\sqrt{\mu/\varepsilon_0}$, the Eq. (3.2.12) become as follows

$$\text{Re}\,\sigma_{\varphi\varphi} = \frac{e^2\omega}{8\pi m_*^2 \left(a\varepsilon_0 |m|\right)^3}\begin{cases}(k_+ - k_-), & 0 < \omega < \omega_-,\\ k_+, & \omega_- < \omega < \omega_+,\\ 0, & \omega > \omega_+,\end{cases}$$

(3.2.13)

$$\text{Re}\,\sigma_{zz} = \frac{e^2}{6\pi m_*^2 a\varepsilon_0 |m|\omega}\begin{cases}(k_+^3 - k_-^3), & 0 < \omega < \omega_-,\\ k_+^3, & \omega_- < \omega < \omega_+,\\ 0, & \omega > \omega_+,\end{cases}$$

(3.2.14)

where

$$\omega_{\pm} = 2\varepsilon_0 |m| \sqrt{\frac{\mu}{\varepsilon_0}} \pm \varepsilon_0 m^2 .$$

Equations (3.2.13) and (3.2.14) are

$$\mathrm{Re}\,\sigma_{\varphi\varphi} = \frac{e^2}{8\pi m_*^2 \varepsilon_0^2 \left(a|m|\right)^3 d} G_a ,$$

$$\mathrm{Re}\,\sigma_{zz} = \frac{e^2}{6\pi m_*^2 \varepsilon_0^2 a|m| d^3} G_b ,$$

where

$$G_a(x) = \begin{cases} x \left[\arccos \dfrac{\varepsilon_0 \left(x - \dfrac{\mu}{\varepsilon_0}\right)^2}{4m^2 \Delta} - \arccos \dfrac{\varepsilon_0 \left(x + \dfrac{\mu}{\varepsilon_0}\right)^2}{4m^2 \Delta} \right], & 0 < x < x_-, \\[4em] x \arccos \dfrac{\varepsilon_0 \left(x - \dfrac{\mu}{\varepsilon_0}\right)^2}{4m^2 \Delta}, & x_- < x < x_+, \\[4em] 0, & x > x_+, \end{cases}$$

$$G_b(x) = \begin{cases} \dfrac{1}{x}\left[\left(\arccos\dfrac{\varepsilon_0\left(x-\dfrac{\mu}{\varepsilon_0}\right)^2}{4m^2\Delta}\right)^3 - \left(\arccos\dfrac{\varepsilon_0\left(x+\dfrac{\mu}{\varepsilon_0}\right)^2}{4m^2\Delta}\right)^3\right], & 0 < x < x_-, \\[2em] \dfrac{1}{x}\left(\arccos\dfrac{\varepsilon_0\left(x-\dfrac{\mu}{\varepsilon_0}\right)^2}{4m^2\Delta}\right)^3, & x_- < x < x_+, \\[2em] 0, & x > x_+, \end{cases}$$

$x = \omega/\varepsilon_0$, $x_\pm = \omega_\pm/\varepsilon_0$. The value $(\mu/\varepsilon_0)^{1/2}$ in $\omega_\pm$ is equal to the classical angular moment ak_F of Fermi electron in the circular orbit (k_F is Fermi momentum).

In Figure 9 (a, b) the dependences of G_a and G_b functions on x are shown for parameters $m = 10$, $\mu/\varepsilon_0 = \Delta/\varepsilon_0 = 100$ typical for semiconductor superlattices.

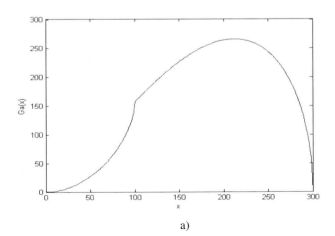

a)

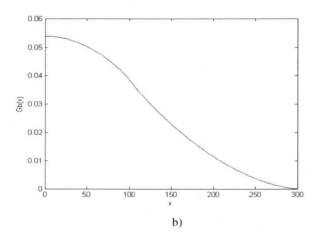

b)

Figure 9. The real part of conductivity (3.2.13) and (3.2.14) depending on frequency for values of parameters referred in the text under condition of $|m| < 2\sqrt{\mu/\varepsilon_0}$.

At $|m| > 2\sqrt{\mu/\varepsilon_0}$ we obtain

$$\mathrm{Re}\,\sigma_{\varphi\varphi} = \frac{e^2 \omega}{8\pi m_*^2 \left(a\varepsilon_0 |m|\right)^3} \begin{cases} 0, & 0 < \omega < \omega_-, \\ k_+, & \omega_- < \omega < \omega_+, \\ 0, & \omega > \omega_+, \end{cases} \qquad (3.2.15)$$

$$\mathrm{Re}\,\sigma_{zz} = \frac{e^2}{6\pi m_*^2 a\varepsilon_0 |m|\omega} \begin{cases} 0, & 0 < \omega < \omega_-, \\ k_+^3, & \omega_- < \omega < \omega_+, \\ 0, & \omega > \omega_+, \end{cases} \qquad (3.2.16)$$

where now $\omega_\pm = \pm 2\varepsilon_0 |m|\sqrt{\dfrac{\mu}{\varepsilon_0}} + \varepsilon_0 m^2$.

Functions (3.2.15) and (3.2.16) are represented as

$$\mathrm{Re}\,\sigma_{\varphi\varphi} = \frac{e^2}{8\pi m_*^2 \varepsilon_0^2 \left(a|m|\right)^3 d} F_a,$$

$$\mathrm{Re}\,\sigma_{zz} = \frac{e^2}{6\pi m_*^2 \varepsilon_0^2 a |m| d^3} F_b,$$

where

$$F_a(x) = x \arccos \frac{\varepsilon_0 (x - m^2)^2}{4m^2 \Delta}, \quad x_- < x < x_+,$$

$$F_b(x) = \frac{1}{x} \left[\arccos \frac{\varepsilon_0 (x - m^2)^2}{4m^2 \Delta} \right]^3, \quad x_- < x < x_+,$$

$x = \omega/\varepsilon_0$, $x_\pm = \omega_\pm/\varepsilon_0$.

In Figure 10 (a, b) the dependences of F_a and F_b functions on x are shown for parameters $m = 5$, $\mu/\varepsilon_0 = \Delta/\varepsilon_0 = 4$ under condition of $|m| > 2\sqrt{\mu/\varepsilon_0}$

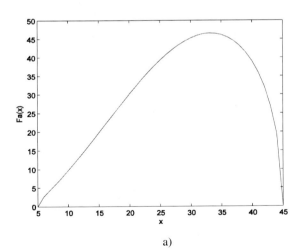

a)

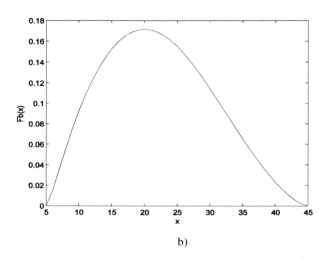

b)

Figure 10. The real part of conductivity from Eqs. (3.2.15) and (3.2.16) depending on frequency for values of parameters referred in the text under condition of $|m| > 2\sqrt{\mu/\varepsilon_0}$.

As the real part of conductivity is connected with electromagnetic field energy absorbed by electrons, the Eqs. (3.2.13)-(3.2.16) determine the boundaries of Landau damping for waves with positive and negative helicity. These boundaries are parabolas in the "angular moment – frequency" plane. The real part of the conductivity determines the damping decrement of electromagnetic waves on the tube.

High-frequency asymptotics of the conductivity imaginary part at $\omega \gg \Omega_\pm$ are

$$\mathrm{Im}\,\sigma_{\varphi\varphi} = \frac{e^2 n}{m_* \omega} + \frac{2e^2 m}{\pi^2 m_*^2 a^3 \omega^2} \sum_l k_l (l+\eta), \qquad (3.2.17)$$

$$\mathrm{Im}\,\sigma_{zz} = \frac{e^2 n}{m_* \omega} + \frac{2e^2 \varepsilon_0 m^2}{3\pi^2 m_*^2 a \omega^3} \sum_l k_l^3. \qquad (3.2.18)$$

If $\varepsilon_0 \ll \mu$, the sums by l included in Eqs. (3.2.17) and (3.2.18) are calculated by Poisson formula:

$$\sum_{l=-\infty}^{\infty} \varphi(l) = \sum_{r=-\infty}^{\infty} \int_{-\infty}^{\infty} dx\, \varphi(x) \exp(2\pi i r x).$$

The limits of integration we derive from the condition

$$\left| \varepsilon_0 (x+\eta)^2 + \Delta - \mu \right| \le \Delta.$$

If we have $\mu < 2\Delta$, then

$$\sum_l \to 2 \int_0^{-\eta+\sqrt{\frac{\mu}{\varepsilon_0}}} dx.$$

If $\mu > 2\Delta$, then

$$\sum_l \to 2 \int_{-\eta+\sqrt{\frac{\mu-2\Delta}{\varepsilon_0}}}^{-\eta+\sqrt{\frac{\mu}{\varepsilon_0}}} dx.$$

Then the Eqs. (3.2.17) and (3.2.18) components contain monotonic $\left(\sigma^{mon} \right)$ addends and oscillating $\left(\sigma^{osc} \right)$ ones. These depend on the ratio of Fermi energy to the miniband width. When $\omega \gg \Omega_\pm$, $\varepsilon_0 \ll \mu$ and $\mu < 2\Delta$ after integration by parts and replacing the integration variable, the oscillating part of the sum

$$J = \sum_l k_l (l+\eta)$$

in formula (3.2.17) is equal to

$$J_{osc} = -\frac{4}{\pi d}\sqrt{\frac{\mu}{\varepsilon_0}}\sum_{r=1}^{\infty}\frac{1}{r}\sin 2\pi r\eta\int_0^1 dy\,\frac{y^2\cos\left(2\pi r\sqrt{\frac{\mu}{\varepsilon_0}}\,y\right)}{\sqrt{\left(1-y^2\right)\left(y^2+\alpha^2\right)}},$$

where

$$\alpha = \left(\frac{2\Delta-\mu}{\mu}\right)^{1/2}.$$

Asymptotic of this integral under $\mu \gg \varepsilon_0$ as is known [87]. As a result, the imaginary part of the transverse conductivity (3.2.17) is equal

$$\mathrm{Im}\,\sigma_{\varphi\varphi} = \frac{e^2 n}{m_* \omega} - \frac{2^{3/2}e^2 m}{\pi^3 m_*^2 a^3 d\omega^2}\left(\frac{\mu}{\varepsilon_0}\right)^{1/4}\times$$

$$\times\left(\frac{\mu}{\Delta}\right)^{1/2}\sum_{r=1}^{\infty}\frac{\sin 2\pi r\eta}{r^{3/2}}\cos\left(2\pi r\sqrt{\frac{\mu}{\varepsilon_0}}-\frac{\pi}{4}\right). \tag{3.2.19}$$

In the case of $\mu > 2\Delta$ we obtain

$$\mathrm{Im}\,\sigma_{\varphi\varphi} = \frac{e^2 n}{m_* \omega} - \frac{2^{3/2}e^2 m}{\pi^3 m_*^2 a^3 d\omega^2}\left(\frac{\mu}{\varepsilon_0}\right)^{1/4}\left(\frac{\mu}{\Delta}\right)^{1/2}\sum_{r=1}^{\infty}\frac{\sin 2\pi r\eta}{r^{3/2}}\times$$

$$\times\left[\cos\left(2\pi r\sqrt{\frac{\mu}{\varepsilon_0}}-\frac{\pi}{4}\right)+\left(\frac{\mu-2\Delta}{\mu}\right)^{3/4}\cos\left(2\pi r\sqrt{\frac{\mu-2\Delta}{\varepsilon_0}}+\frac{\pi}{4}\right)\right]. \tag{3.2.20}$$

The Eqs. (3.2.19) and (3.2.20) undergoes Aharonov-Bohm oscillations under variation of magnetic flux through the tube cross-section. The oscillation period is equal to the flux quantum Φ_0. Also the oscillations looking like de Haas-van Alphen ones exist. They are caused by transition of root singularities of electron density of states at the miniband boundaries through Fermi boundary due to the tube radius variation or changing the electron density. The latter is related with Fermi energy as follows

$$\mu = \begin{cases} \dfrac{\pi}{2}dn\sqrt{\dfrac{\Delta}{m_*}}, & \mu << 2\Delta, \\[3mm] \dfrac{1}{8m_*}(\pi dn)^2, & \mu >> 2\Delta. \end{cases}$$

Analyzing the dependence of oscillations (3.2.19) on $(adn)^{1/2}$ we obtain the period $\tau = \left(\dfrac{1}{\pi a\sqrt{m_*\Delta}}\right)^{1/2}$.

If $\mu < 2\Delta$, only the miniband bottom boundaries ε_l pass through Fermi boundary when the tube parameters change. As a result, in Eq. (3.2.19) the base frequency of oscillations is present only. The second addend in Eq. (3.2.20) exists because at $\mu > 2\Delta$ not only miniband ε_l bottom boundaries transverse Fermi level but the upper ones $\varepsilon_l + 2\Delta$ as well. Existence of two oscillation frequencies in Eq. (3.2.20) causes the beats in the plot of conductivity versus the tube parameters. They are similar to the beats of plasma and spin waves spectra in the tube [86, 87]. If $\Delta \ll \mu$ the relative difference of conductivity oscillation frequencies and amplitudes (3.2.20) is of the order of Δ/μ. As this parameter increases, the beats turn into weak modulations and disappear at $\mu < 2\Delta$.

The sum by l in longitudinal conductivity (3.2.18) is calculated by Poisson formula as well. Consequently, the conductivity contains monotonic and oscillating components. At $\mu > 2\Delta$ they are equals

$$\operatorname{Im} \sigma_{zz}^{mon} = \frac{e^2 n}{m_* \omega} + \frac{2e^2 \varepsilon_0 m^2}{3\pi^2 m_*^2 a \omega^3 d^3} J_{mon}, \quad (3.2.21)$$

$$\operatorname{Im} \sigma_{zz}^{osc} = \frac{2e^2 \varepsilon_0 m^2}{3\pi^2 m_*^2 a \omega^3 d^3} \sum_{r=1}^{\infty} \cos 2\pi r \eta \cdot J_{osc}^r, \quad (3.2.22)$$

where

$$J_{mon}(b) = 2 \int_{\sqrt{b^2 - 2c^2}}^{b} dx \left(\arccos \frac{x^2 - b^2 + c^2}{c^2} \right)^3, \quad (3.2.23)$$

$$J_{osc}^r(b) = 4 \int_{\sqrt{b^2 - 2c^2}}^{b} dx \cos(2\pi r x) \left(\arccos \frac{x^2 - b^2 + c^2}{c^2} \right)^3, \quad (3.2.24)$$

$b = (\mu/\varepsilon_0)^{1/2}$, $c^2 = \Delta/\varepsilon_0$.

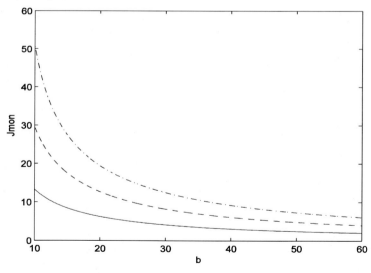

a)

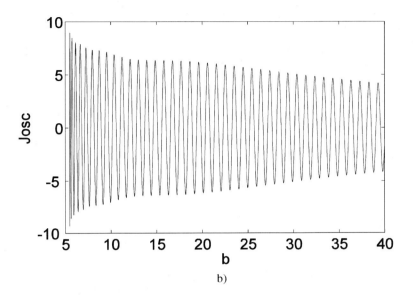

b)

Figure 11. Monotonic and oscillating components of longitudinal conductivity (3.2.21) and (3.2.22) at $\mu > 2\Delta$ as the functions of $(\mu/\varepsilon_0)^{1/2}$ for parameter values referred in the text.

The integrals (3.2.23) and (3.2.24) are not calculated exactly. In Fig. 11 (a), the dependence $J_{mon}(b)$ calculated numerically is shown. Solid-, dotted-, and chain-line curves correspond to $c^2 = 10, 20, 30$, respectively. The $J_{osc}(b)$ dependence at $r = 1$ and $c^2 = 15$ is shown in Fig. 11 (b).

In accordance with formulas (3.2.21)-(3.2.24) the monotonic part of the conductivity and the amplitude of the oscillating part decrease with frequency ω_n increases. Fig. 11 (b) shows the weak modulations are caused by the beats.

In accordance with formulas (3.2.21)-(3.2.24) the monotonic part of the conductivity and the amplitude of the oscillating part decrease with frequency ω_n increases. Fig. 11 (b) shows the weak modulations are caused by the beats.

The imaginary part of the transverse conductivity (3.2.19) and (3.2.20) behaves similarly. It is includes into the dispersion equation for the electromagnetic waves spectrum. The beats and oscillations obtained here there exist only in the quasiclassical case. Their reasons were described above.

In the absence of a superlattice from Eqs. (3.2.18) and (3.2.19) we obtain [62]

$$\mathrm{Im}\,\sigma_{\varphi\varphi}^{osc} = -\frac{2e^2 m}{\pi^3 m_*^2 a^4 \omega^2}\left(\frac{\mu}{\varepsilon_0}\right)^{3/4}\sum_{r=1}^{\infty}\frac{\sin 2\pi r\eta}{r^{3/2}}\cos\left(2\pi r\sqrt{\frac{\mu}{\varepsilon_0}}-\frac{\pi}{4}\right),$$

$$\mathrm{Im}\,\sigma_{zz}^{mon} = \frac{e^2 n}{m_* \omega}+\frac{e^2 \varepsilon_0 m^2}{2\pi m_*^2 a^4 \omega^3}\left(\frac{\mu}{\varepsilon_0}\right)^2,$$

$$\mathrm{Im}\,\sigma_{zz}^{osc} = -\frac{e^2 \varepsilon_0 m^2}{\pi^3 m_*^2 a^4 \omega^3}\left(\frac{\mu}{\varepsilon_0}\right)^{3/4}\sum_{r=1}^{\infty}\frac{\cos 2\pi r\eta}{r^{5/2}}\cos\left(2\pi r\sqrt{\frac{\mu}{\varepsilon_0}}-\frac{\pi}{4}\right).$$

Here we have used $k_l = \left[2m_*(\mu-\varepsilon_l)\right]^{1/2}$. In the Poisson formula the asymptotic of the table integral was used [88]

$$\int_0^a dx \cos bx\left(a^2 - x^2\right)^{\beta-1} = \frac{\sqrt{\pi}}{2}\left(\frac{2a}{b}\right)^{\beta-\frac{1}{2}}\Gamma(\beta)J_{\beta-\frac{1}{2}}(ab),$$

where $\Gamma(\beta)$ is the Euler function, $J_{\beta-1/2}(ab)$ is the Bessel function at $ab \gg 1$. The ratio of the oscillation amplitude of the transverse conductivity (3.2.19) to the amplitude of the oscillation in the absence of the superlattice is of order $a/d\left(\varepsilon_0/\Delta\right)^{1/2}$. The longitudinal conductivity oscillations of de Haas-van Alphen type exist also in the absence of magnetic field.

3.3. Spatial Dispersion of the Conductivity of the Electron Gas on the Surface of a Nanotube with a Superlattice in a Magnetic Field

The electron energy on the surface of a cylindrical nanotube with a longitudinal superlattice in a longitudinal magnetic field has the form [37, 61, 85, 86, 89, 90]

$$\varepsilon_{lk} = \varepsilon_0 \left(l + \eta\right)^2 + \Delta\left(1 - \cos kd\right), \tag{3.3.1}$$

where l and k are the projections of angular momentum and electron momentum on the tube axis z, $\varepsilon_0 = \left(2m_* a^2\right)^{-1}$ is the rotational quantum, m_* is the effective electron mass, a is the tube radius, η is the ratio of the magnetic flux through the tube to the quantum flux [74], Δ and d are the amplitude and the period of the modulating potential acting on the electron. The quantum constant is taken equal to one. Spin level splitting is not taken into account. The second term on the right-hand side of formula (3.3.1) is borrowed from the theory of the strong coupling of conduction electrons to the lattice.

Reaction of an electron gas to a weak electric field

$$\vec{E} = \vec{E}_0 \exp i\left(m\varphi + qz - \omega t\right) \tag{3.3.2}$$

of electromagnetic wave is characterized by a two-dimensional surface conductivity tensor $\sigma_{\alpha\beta}\left(m, q, \omega\right)$. Here m and q are the projections of angular momentum and wave momentum on the tube axis, ω is the wave frequency, φ and z are the cylindrical coordinates. The density of the surface current on the tube is equal to

$$j_\alpha(m,q,\omega) = \sum_\beta \sigma_{\alpha\beta}(m,q,\omega) E_\beta(m,q,\omega). \qquad (3.3.3)$$

The indexes m,q denote the cylindrical harmonics of the values included in the formula (3.3.3).

Using the theory of linear reaction [83], we obtain the Kubo formula for the conductivity tensor:

$$\sigma_{\alpha\beta}(m,q,\omega) = i\frac{e^2 n}{m_*\omega}\delta_{\alpha\beta} + \frac{1}{\omega}\int_0^\infty dt\, e^{i\omega t}\left\langle\left[J_\alpha(m,q,t), J_\beta(-m,-q,0)\right]\right\rangle, \qquad (3.3.4)$$

where e is the electron charge, n is the surface density of electrons, $\vec{J}(m,q,t)$ is the cylindrical harmonic current density operator. Angle brackets denote the Gibbs average of the operator commutator.

The cylindrical components of the conductivity tensor are:

$$\sigma_{\varphi\varphi} = i\frac{e^2 n}{m_*\omega} + i\frac{2e^2}{m_*^2 a^2 \omega S}\sum_{lk} f(\varepsilon_{lk}) \times$$

$$\times\left[\frac{\left(l+\eta+\dfrac{m}{2}\right)^2}{\varepsilon_{lk}-\varepsilon_{(l+m)(k+q)}+\omega+i0} - \frac{\left(l+\eta-\dfrac{m}{2}\right)^2}{\varepsilon_{(l-m)(k-q)}-\varepsilon_{lk}+\omega+i0}\right],$$

$$\sigma_{\varphi z} = \sigma_{z\varphi} = i\frac{2e^2}{m_*^2 a\omega S}\sum_{lk} f(\varepsilon_{lk}) \times$$

$$\times\left[\frac{\left(l+\eta+\dfrac{m}{2}\right)\left(k+\dfrac{q}{2}\right)}{\varepsilon_{lk}-\varepsilon_{(l+m)(k+q)}+\omega+i0} - \frac{\left(l+\eta-\dfrac{m}{2}\right)\left(k-\dfrac{q}{2}\right)}{\varepsilon_{(l-m)(k-q)}-\varepsilon_{lk}+\omega+i0}\right], \qquad (3.3.5)$$

Electron Gas on the Surface of a Nanotube

$$\sigma_{zz} = i\frac{e^2 n}{m_* \omega} + i\frac{2e^2}{m_*^2 \omega S}\sum_{lk} f(\varepsilon_{lk}) \times$$

$$\times \left[\frac{\left(k+\dfrac{q}{2}\right)^2}{\varepsilon_{lk} - \varepsilon_{(l+m)(k+q)} + \omega + i0} - \frac{\left(k-\dfrac{q}{2}\right)^2}{\varepsilon_{(l-m)(k-q)} - \varepsilon_{lk} + \omega + i0} \right].$$

Here f is the Fermi function, $S = 2\pi aL$ is the lateral surface area of tube length L.

In the absence of spatial dispersion from formulas (3.3.5) at zero temperature for a degenerate electron gas we obtain

$$\mathrm{Re}\,\sigma_{\varphi\varphi}(m,\omega) = \frac{e^2}{\pi m_*^2 a^3 \omega}\sum_l k_l \times$$

$$\times \left[\left(l+\eta+\frac{m}{2}\right)^2 \delta(\omega-\Omega_+) - \left(l+\eta-\frac{m}{2}\right)^2 \delta(\omega-\Omega_-) \right], \qquad (3.3.6)$$

$$\mathrm{Im}\,\sigma_{\varphi\varphi} = \frac{e^2 n}{m_* \omega} + \frac{e^2}{\pi^2 m_*^2 a^3 \omega}\sum_l k_l \left[\frac{\left(l+\eta+\dfrac{m}{2}\right)^2}{\omega-\Omega_+} - \frac{\left(l+\eta-\dfrac{m}{2}\right)^2}{\omega-\Omega_-} \right],$$

$$\sigma_{\varphi z}(m,\omega) = 0,$$

$$\mathrm{Re}\,\sigma_{zz}(m,\omega) = \frac{e^2}{3\pi m_*^2 a\omega}\sum_l k_l^3 \left[\delta(\omega-\Omega_+) - \delta(\omega-\Omega_-) \right],$$

$$\mathrm{Im}\,\sigma_{zz}(m,\omega) = \frac{e^2 n}{m_* \omega} + \frac{e^2}{3\pi^2 m_*^2 a\omega}\sum_l k_l^3 \left[\frac{1}{\omega-\Omega_+} - \frac{1}{\omega-\Omega_-} \right],$$

where

$$k_l = \frac{1}{d}\arccos\frac{\varepsilon_l + \Delta - \mu}{\Delta} \tag{3.3.7}$$

is the maximum electron momentum in a miniband with a number l, $\varepsilon_l = \varepsilon_0(l+\eta)^2$ is the bottom boundary of the miniband, μ is the Fermi energy,

$$\Omega_\pm = \varepsilon_0 m\left[2(l+\eta)\pm m\right] \tag{3.3.8}$$

are the frequencies of direct transitions of electrons in the field of an electromagnetic wave between minibands. Summation over l in (3.3.6) is limited by the condition $|\varepsilon_l + \Delta - \mu| \le \Delta$. It means that the Fermi level is located in the miniband.

In the case $q = 0$ the intraband electron transitions are absent. The interband current only remains. If $m = 0$, the real part of conductivity (3.3.6) as expected is equal to zero. Since it determines the energy of an electromagnetic wave absorbed by electrons, there is no absorption in the absence of direct and indirect electron transitions. The surface diamagnetic current only remains on the tube.

The real part of the transverse and longitudinal conductivity (3.3.6) as a function of frequency has narrow maxima at the frequencies of direct transitions of electrons between minibands. The imaginary part has resonant peculiarities at these frequencies. The oscillator strengths of resonant electron transitions are determined by the maximum electron momentum in partially filled minibands. Superlattice parameters Δ and d enter into (3.3.6) only through these momenta. In particular, at $\eta < 1/2$ the lower boundaries of minibands ε_l satisfy inequalities $\varepsilon_0\eta^2 < \varepsilon_{-1} < \varepsilon_1 < \varepsilon_{-2} < ...$ If the two bottom minibands with widths 2Δ overlap so that $\varepsilon_0\eta^2 + 2\Delta > \varepsilon_{-1}$ and the Fermi level is located in the

Electron Gas on the Surface of a Nanotube 79

overlap region $\left[\varepsilon_{-1}, \varepsilon_0 \eta^2 + 2\Delta\right]$, then the oscillator strengths are determined by the values k_0 and k_{-1}. If there is no overlap of the minibands, the maximum electron momentum in the fully filled lower miniband coincides with the Brillouin zone boundary π/d.

Taking into account the spatial dispersion, the real part of the conductivity tensor depends on the position of the Fermi level in the miniband. If it is located in the lower half of the miniband with the number l, i.e., $\varepsilon_l < \mu < \varepsilon_l + \Delta$, $0 < q < \pi/2d$, we obtained

$$\operatorname{Re}\sigma_{zz}(m,q,\omega) = \frac{e^2}{2\pi m_*^2 ad\omega} \sum_l \frac{\left(k_l^-\right)^2}{\left[4\Delta^2 \sin^2 \frac{qd}{2} - (\omega - \Omega_+)^2\right]^{\frac{1}{2}}}, \quad (3.3.9)$$

where

$$k_l^- = \frac{1}{d} \arcsin \frac{\omega - \Omega_+}{2\Delta \sin \frac{qd}{2}}, \qquad \omega_- < \omega < \omega_+,$$

$$\omega_\pm = \Omega_+ \pm 2\Delta \sin \frac{qd}{2}\left(\alpha_l \sin \frac{qd}{2} + \sqrt{1 - \alpha_l^2}\cos \frac{qd}{2}\right),$$

$$\alpha_l = \frac{\varepsilon_l + \Delta - \mu}{\Delta}. \tag{3.3.10}$$

If $\varepsilon_l + \Delta < \mu < \varepsilon_l + 2\Delta$, $\pi/2d < q < \pi/d$, we obtaine

$$\operatorname{Re}\sigma_{zz}(m,q,\omega) = \frac{e^2}{2\pi m_*^2 ad^3\omega} \sum_l \frac{\left(\pi - dk_l^-\right)^2}{\left[4\Delta^2 \sin^2 \frac{qd}{2} - (\omega - \Omega_+)^2\right]^{\frac{1}{2}}}, \quad (3.3.11)$$

where $\omega_- < \omega < \omega_+$,

$$\omega_\pm = \Omega_+ \pm 2\Delta \sin\frac{qd}{2}\left(|\alpha_l|\sin\frac{qd}{2} + \sqrt{1-\alpha_l^2}\cos\frac{qd}{2}\right). \quad (3.3.12)$$

The remaining components $\text{Re}\,\sigma_{\alpha\beta}$ can be obtained by replacing in the formula (3.3.9) $\left(k_l^-\right)^2$ by $k_l^-(l+\eta+m/2)/a$ in the expression for $\text{Re}\,\sigma_{\varphi z}$ and by $(l+\eta+m/2)^2/a^2$ in $\text{Re}\,\sigma_{\varphi\varphi}$. Replacing in the same way $\left(\pi - dk_l^-\right)^2$ in the formula (3.3.11) we obtain $\text{Re}\,\sigma_{\varphi z}$ and $\text{Re}\,\sigma_{\varphi\varphi}$ in the upper half of the miniband

From formulas (3.3.9)-(3.3.12) it follows that the real part of the conductivity of a degenerate electron gas is nonzero in the Landau damping region on the plane $q-\omega$. It is schematically shown in Figure 12. In this area, electromagnetic waves propagating along the tube undergoes collisionless

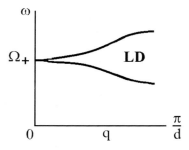

Figure 12. Landau damping region of electromagnetic waves on a tube with a superlattice.

Landau damping (LD)

If the inequality $\varepsilon_0 \ll \mu$ holds, the quantization of the transverse motion of electrons on the tube can be neglected. Replacing in this case the summation of l in formulas (3.3.6) by integrals in the absence of spatial dispersion at $|m| < 2\sqrt{\mu/\varepsilon_0}$, we obtained

Electron Gas on the Surface of a Nanotube

$$\operatorname{Re}\sigma_{\varphi\varphi} = \frac{e^2\omega}{8\pi m_*^2 \left(a\varepsilon_0 |m|\right)^3} \begin{cases} \left(k_+ - k_-\right), & 0 < \omega < \omega_-, \\ k_+, & \omega_- < \omega < \omega_+, \\ 0, & \omega > \omega_+, \end{cases} \qquad (3.3.13)$$

$$\operatorname{Re}\sigma_{zz} = \frac{e^2}{6\pi m_*^2 a\varepsilon_0 |m|\omega} \begin{cases} \left(k_+^3 - k_-^3\right), & 0 < \omega < \omega_-, \\ k_+^3, & \omega_- < \omega < \omega_+, \\ 0, & \omega > \omega_+. \end{cases} \qquad (3.3.14)$$

Here

$$k_{\pm} = \frac{1}{d}\arccos\frac{\varepsilon_{\pm} + \Delta - \mu}{\Delta}, \, \varepsilon_{\pm} = \varepsilon_0 \left(\frac{\omega}{2m\varepsilon_0} \mp \frac{m}{2}\right)^2, \qquad (3.3.15)$$

$$\omega_{\pm} = 2\varepsilon_0 |m|\sqrt{\frac{\mu}{\varepsilon_0}} \pm \varepsilon_0 m^2.$$

Magnitude $\left(\mu/\varepsilon_0\right)^{1/2}$ entering into $\omega_{\pm}$ is equal to the classical angular momentum ak_F of Fermi electron in a circular orbit (k_F is the Fermi momentum). Formulas (3.3.15) determine the Landau damping boundaries on the plane angular momentum-frequency.

The imaginary part of the conductivity tensor is included in the dispersion equation for the spectrum of electromagnetic waves propagating along the tube. In the absence of spatial dispersion, it is equal to (3.3.6). Accounting for weak spatial dispersion if $\Delta\sin qd/2 << |\omega - \Omega_{\pm}|$ leads to corrections $\operatorname{Im}\delta\sigma$ to formulas (3.3.6) proportional to q and q^2.

The imaginary part of the transverse conductivity of a degenerate electron gas contains the correction

$$\text{Im}\,\delta\sigma_{\varphi\varphi}(m,q,\omega)=\frac{2e^2\Delta}{\pi^2 m_*^2 a^3 d\omega}\sin^2\frac{qd}{2}\sum_l \sin k_l d\left[\frac{\left(l+\eta+\dfrac{m}{2}\right)^2}{\left(\omega-\Omega_+\right)^2}+\frac{\left(l+\eta-\dfrac{m}{2}\right)^2}{\left(\omega-\Omega_-\right)^2}\right].\qquad (3.3.16)$$

In the absence of interband transitions $\left(m=0\right)$ we obtained

$$\text{Im}\,\delta\sigma_{\varphi\varphi}(q,\omega)=\frac{4e^2\sin^2\dfrac{qd}{2}}{\pi^2 m_*^2 a^3 d\omega^3}\sum_l (l+\eta)^2\left[\left(\mu-\varepsilon_l\right)\left(\varepsilon_l+2\Delta-\mu\right)\right]^{1/2}.$$

In the limits $d\to 0$, $\Delta\to\infty$, $d\sqrt{\Delta}\to m_*^{-1/2}$ it will take the form of the transverse conductivity of a nanotube in the absence of a superlattice due to intraband electron transitions [62]

$$\text{Im}\,\delta\sigma_{\varphi\varphi}(q,\omega)=\frac{e^2 q^2}{\pi^2\left(m_* a\omega\right)^3}\sum_l (l+\eta)^2 k_l,\qquad (3.3.17)$$

where

$$k_l=\sqrt{2m_*\left(\mu-\varepsilon_l\right)}.$$

In the case $m=0$ and $\varepsilon_0\ll\mu$ in the absence of a superlattice from formula (3.3.17) we obtain

$$\text{Im}\,\delta\sigma_{\varphi\varphi}(q,\omega)=\frac{e^2\mu^2 q^2}{2\pi m_*\omega^3}\left[1+\frac{8}{\pi^2}\left(\frac{\varepsilon_0}{\mu}\right)^{3/4}\sum_{r=1}^{\infty}\frac{1}{r^{3/2}}\cos 2\pi r\eta\cdot\sin\left(2\pi r\sqrt{\frac{\mu}{\varepsilon_0}}-\frac{\pi}{4}\right)\right].\quad (3.3.18)$$

Electron Gas on the Surface of a Nanotube

This expression undergoes Aaronov-Bohm oscillations and de Haas-van Alphen-type oscillations similar to the oscillations of the spectrum of magnetoplasma and spin waves [61, 86]. Oscillations of the de Haas-van Alphen type are also preserved in the absence of a magnetic field. They are due to the passage of the root features of the density of states of electrons at the points ε_l through the Fermi level with nanotube parameters changes.

The non-diagonal component of the conductivity tensor in the long-wavelength limit at $k_l d \ll 1$ equals to

$$\operatorname{Im}\sigma_{\varphi z}(m,q,\omega) = \frac{2e^2 d\Delta\sin\frac{qd}{2}}{3\pi^2 m_*^2 a^2 \omega} \sum_l k_l^3 \left[\frac{l+\eta+\frac{m}{2}}{(\omega-\Omega_+)^2} - \frac{l+\eta-\frac{m}{2}}{(\omega-\Omega_-)^2} \right].$$

In the absence of a superlattice from here we obtain

$$\operatorname{Im}\sigma_{\varphi z}(m,q,\omega) = \frac{e^2 q}{3\pi^2 m_*^3 a^2 \omega} \sum_l k_l^3 \left[\frac{l+\eta+\frac{m}{2}}{(\omega-\Omega_+)^2} - \frac{l+\eta-\frac{m}{2}}{(\omega-\Omega_-)^2} \right].$$

The corrections to the longitudinal conductivity (3.3.6) at $\Delta\sin qd/2 \ll |\omega-\Omega_\pm|$ and $qd \ll 1$ equals to

$$\operatorname{Im}\delta\sigma_{zz}(m,q,\omega) = \frac{e^2 d\Delta q^2}{2\pi^2 m_*^2 a\omega} \sum_l k_l^2 \sin k_l d \left[\frac{1}{(\omega-\Omega_+)^2} + \frac{1}{(\omega-\Omega_-)^2} \right].$$

In the absence of a superlattice, the expression follows:

$$\operatorname{Im}\delta\sigma_{zz}(m,q,\omega) = \frac{e^2 q^2}{2\pi^2 m_*^3 a\omega} \sum_l k_l^3 \left[\frac{1}{(\omega-\Omega_+)^2} + \frac{1}{(\omega-\Omega_-)^2} \right].$$

If here $m = 0$, $\varepsilon_0 \ll \mu$ we obtain the correction to the longitudinal conductivity (3.3.6) in the semiclassical case:

$$\operatorname{Im}\delta\sigma_{zz}(q,\omega) = \frac{3e^2\mu^2 q^2}{2\pi m_*\omega^3}\left[1 + \frac{4}{\pi^2}\frac{\varepsilon_0}{\mu}\sum_{r=1}^{\infty}\frac{1}{r^2}\cos 2\pi r\eta \cdot \mathrm{J}_2\left(2\pi r\sqrt{\frac{\mu}{\varepsilon_0}}\right)\right], \quad (3.3.19)$$

where J_2 is the Bessel function. As the transverse conductivity, the longitudinal conductivity (3.3.19) undergoes Aharonov-Bohm oscillations with a change of the magnetic flux through the tube section and oscillations of the de Haas-van Alphen type with a change of the tube radius or electron density.

Conclusion

Kubo formula was derived in the Subsection 3.1 for the electron gas conductivity tensor on the nanotube surface in longitudinal magnetic field considering spatial and time dispersion. Components of the degenerate and nondegenerate electron gas conductivity tensor were calculated. The study has showed that under high electron density, the conductivity undergoes oscillations of de Haas - van Alphen and Aharonov-Bohm types with the density of electrons and magnetic field changes.

Kubo formula was obtained in the Subsection 3.2 for conductivity tensor of electron gas on the surface of nanotube with superlattice in magnetic field. The high-frequency conductivity tensor components were calculated for quantum and quasiclassical cases. Electromagnetic wave Landau damping areas in the tube were determined. The conductivity tensor components show Aharonov-Bohm type oscillations and de Haas-van Alphen ones. When Fermi energy exceeds the miniband width, beats are observed in the plot of conductivity versus the tube parameters. Otherwise, the beats are absent.

High-frequency conductivity tensor of the electron gas on the nanotube surface with a superlattice in a magnetic field is calculated taking into

Electron Gas on the Surface of a Nanotube 85

account the spatial dispersion in the Subsection 3.3. The real part of the conductivity as a function of frequency has narrow peaks at frequencies of electron transitions between minibands.

The imaginary part has a resonant peculiarities at these frequencies. The transparency windows for the electromagnetic waves propagating along the tube has been found. As in the classical case, the conductivity undergoes Aharonov-Bohm oscillations and de Haas-van Alphen oscillations with the magnetic flux through the cross section of the tube and its parameters changes.

4. COLLECTIVE EXCITATIONS OF ELECTRON GAS ON THE SURFACE OF NANOTUBES

4.1. Plasma Waves on the Surface of Nanotube with Superlattice

A review of articles on collective excitations of electron gas on the tube for the period up to 2014 is contained in the collection of articles [58].

Plasma waves on the surface of carbon [1] and semiconductor [92, 93] nanotubes were studied in [37-40, 89, 91, 94]. Plasma waves in the nanotubes are studied mainly in approximation of random phases and in the hydrodynamic approximation. In the framework of the hydrodynamic approach, using the continuity equation for electron liquid and Poisson equation for electrical potential, the authors [89, 91] have obtained the dispersion equation for the spectrum of surface plasma waves on the tube:

$$\omega = 4\pi a \left[\frac{m^2}{a^2} \operatorname{Im} \sigma_{\varphi\varphi}(m,\omega) + q^2 \operatorname{Im} \sigma_{zz}(m,\omega) \right] I_m(qa) K_m(qa), \quad (4.1.1)$$

where a is the tube radius; m is the projection of the plasmon angular moment on the tube axis z; q and ω are respectively the wave vector and wave frequency; $\sigma_{\varphi\varphi}$ and σ_{zz} are components of electron gas dynamical conductivity in absentia of spatial dispersion ($qv_0 \ll \omega$, v_0 is the Fermi velocity) in cylindrical coordinates φ, z; I_m and K_m are modified Bessel functions.

The equation (4.1.1) is true also for the tube with a superlattice. It can be obtained by embedding fullerenes or other additives to the nanotube or when the nanotube is attached to a substrate for charge exchange [95-97]. Substituting Drude expression for conductivity $ie^2 n / m_* \omega$ into (4.1.1), we obtain the known spectrum for intraband ($m=0$) and interband ($m \neq 0$) plasmons [37-40, 89, 91, 94]:

$$\Omega_{mq}^2 = \frac{4\pi e^2 an}{m_*}\left[\frac{m^2}{a^2}+q^2\right]I_m(qa)K_m(qa),\qquad(4.1.2)$$

where m_* and e are respectively the effective mass and charge; n is the surface density of electrons.

The Eq. (4.1.2) does not take into account the interband current caused by quantum transitions of electrons in the wave field between the minibands. Taking that into account, the transverse component of dynamical conductivity tensor for electron gas on the tube is [62]

$$\sigma_{\varphi\varphi} = i\frac{e^2 n}{m_* \omega} + i\frac{2e^2}{m_*^2 a^2 \omega S}\times$$
$$\times\sum_{lk}n(\varepsilon_{lk})\left[\left(1+\frac{m}{2}\right)^2(\omega-\Omega_+ +i0)^{-1}-\left(1-\frac{m}{2}\right)^2(\omega-\Omega_- +i0)^{-1}\right],\qquad(4.1.3)$$

where l and k are projections of electron angular moment and momentum on the tube axis; $n(\varepsilon_{lk})$ is the Fermi-Dirac function; $\Omega_\pm = \varepsilon_0 m(2l \pm m)$ are frequencies of direct transitions of electrons between the minibands;

$\varepsilon_0 = \left(2m_* a^2\right)^{-1}$ is the rotational quantum [74]; S is the tube surface area. The quantum constant was taken equal to unity. The longitudinal conductivity σ_{zz} is obtained from (4.1.3) using substitution of $a^{-2}\left(l \pm \dfrac{m}{2}\right)^2$ by k^2 and $\sigma_{\varphi z} = \sigma_{z\varphi} = 0$. In (4.1.3) we apply the electron energy on the surface of the semiconductor nanotube with a superlattice [74, 86]:

$$\varepsilon_{lk} = \varepsilon_0 l^2 + \Delta\left(1 - \cos kd\right),$$

where Δ and d are respectively the amplitude and period of the modulating potential.

The imaginary part of the interband conductivity has resonance singularities at frequencies $\Omega_{\pm}$. The Landau attenuation is concentrated in narrow bands $\delta\omega \sim \Delta qd$ near these frequencies [87].

In formulas for conductivity, we restrict ourselves to the quantum limit where electrons in the degenerated gas occupy partially only the lower miniband $l = 0$ with width 2Δ and their density does not exceed $1/\pi ad$. In this case, the solution of Eq. (4.1.1) is defined by the parameter $\alpha_m = 3m^2 / 4\left(ak_0\right)^2$. That is connected with forces of oscillators for resonance transitions $0 \to m$ of electrons between the minibands. Here

$$k_0 = \frac{1}{d}\arccos\frac{\Delta - \mu}{\Delta}$$

is the maximum momentum of an electron in the miniband with $l = 0$; μ is the Fermi energy.

If $\alpha_m < 1$, there exists a series of branches in the plasmon spectrum

$$\omega_{mq}^2 = \frac{1}{2}\left\{\omega_m^2 + \Omega_{mq}^2 + \left[\left(\omega_m^2 + \Omega_{mq}^2\right)^2 + \right.\right.$$

$$\left.\left. + \frac{16}{3}\left(\frac{a}{m}\right)^4 \omega_m^2 \Omega_{mq}^2 \left(1 - \alpha_m\right)(qk_0)^2 \left(1 + \frac{q^2 a^2}{m^2}\right)^{-1}\right]^{1/2}\right\}, \tag{4.1.4}$$

where $\omega_m = \varepsilon_0 m^2$ are frequencies of electron single-particle transitions $0 \to m$. Figure 13 shows the frequency of the wave (4.1.4) $\omega'_{1q} = \omega_{1q}/\Omega_{10}$ (solid line) and wave (4.1.2) $\Omega'_{1q} = \Omega_{1q}/\Omega_{10}$ (dashed curve) as a function of $x = qa$ for $m = 1$ and $\alpha_1 = 0.75$. Here $\Omega_{10} = \left(2\pi e^2 n/m_* a\right)^{1/2}$ is the limiting frequency for the wave with the spectrum (4.1.2). Parameter values $m_* = 0.64 \cdot 10^{-28} g$ (GaAs), $a = 10^{-7}$ cm, $k_0 a = 1$ are used. Under the condition $\alpha_1 < 1$ the Fermi level lies in the upper half of the miniband. If $\alpha_m > 1$, then two branches are connected with each $0 \to m$ transition:

$$\omega_\pm^2(m,q) = \frac{1}{2}\left\{\omega_m^2 + \Omega_{mq}^2 \pm \left[\left(\omega_m^2 + \Omega_{mq}^2\right)^2 - \right.\right.$$

$$\left.\left. - \frac{16}{3}\left(\frac{a}{m}\right)^4 \omega_m^2 \Omega_{mq}^2 \left(\alpha_m - 1\right)(qk_0)^2 \left(1 + \frac{q^2 a^2}{m^2}\right)^{-1}\right]^{1/2}\right\}. \tag{4.1.5}$$

Electron Gas on the Surface of a Nanotube

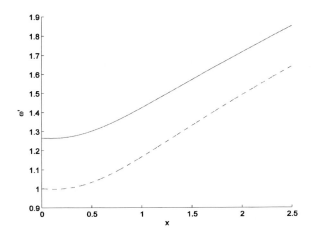

Figure 13. The dispersion curves of waves with the spectrum (4.1.4) (solid line) and with the spectrum (4.1.2) (dashed line) under $m = 1$, $\alpha_1 < 1$ are shown. Parameter values are given in the text.

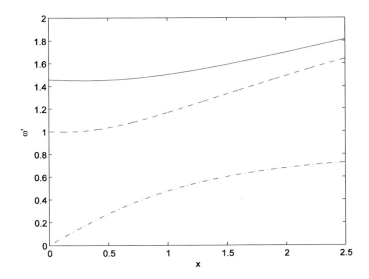

Figure 14. The dispersion curves of waves with the spectrum (4.1.5) (solid and dashed-dotted curves) and with the spectrum (4.1.2) (dashed line) under $m = 1$, $\alpha_1 > 1$ are shown. The parameters values are given in the text.

Figure 14 shows the dependence of the wave frequencies (4.1.5)
$\omega'_{\pm}(1,q) = \omega_{\pm}(1,q) \Big/ \Omega_{10}$ (solid and dashed-dotted curves) and wave

(4.1.2) $\Omega'_{1q} = \Omega_{1q} \Big/ \Omega_{10}$ (dashed curve) as a function of $x = qa$ under

$m = 1$ and $\alpha_1 = 3$. The above mentioned values of m_*, $_a$ and $k_0 a = 0.5$ were used. In this case the Fermi level lies in the lower half of the miniband. The branches (4.1.4) and ω_+ (4.1.5) are positioned above ω_m, and the branches ω_- (4.1.5) are below ω_m.

In the limit of long waves ($qa \ll 1$) and at $\alpha_m < 1$ from Eq. (4.1.4) we obtain

$$\omega_{1q}^2 = \omega_{10}^2 \left[1 + \frac{1}{2} \frac{\Omega_{10}^2}{\omega_1^2 + \Omega_{10}^2} (qa)^2 \ln qa + \right.$$
$$\left. + \frac{4}{3} \frac{\omega_1^2 \Omega_{10}^2}{\left(\omega_1^2 + \Omega_{10}^2 \right)^2} (1 - \alpha_1)(k_0 a)^2 (qa)^2 \right],$$

(4.1.6)

$$\omega_{mq}^2 = \omega_{m0}^2 \left[1 + \frac{\Omega_{m0}^2}{\omega_m^2 + \Omega_{m0}^2} \frac{m^2 - 2}{2m^2 \left(m^2 - 1 \right)} (qa)^2 + \right.$$
$$\left. + \frac{4}{3m^4} \frac{\omega_m^2 \Omega_{m0}^2}{\left(\omega_m^2 + \Omega_{m0}^2 \right)^2} (1 - \alpha_m)(k_0 a)^2 (qa)^2 \right],$$

(4.1.7)

$$m = \pm 2, \pm 3, \dots$$

The critical frequencies of waves with spectra (4.1.6) and (4.1.7) are

$$\omega_{m0}^2 = \omega_m^2 + \Omega_{m0}^2 = \varepsilon_0^2 m^4 + \frac{2e^2 k_0 |m|}{\pi m_* a^2}. \tag{4.1.8}$$

The frequency depolarization shift in (4.1.8) contains the period and the amplitude of the superlattice modulating potential.

At $\alpha_m > 1$ the expressions (4.1.6) and (4.1.7) are true for the upper branch ω_+. The bottom branch ω_- has the sound spectrum $\omega_-(m, q) = c_m q$, where

$$c_m^2 = \frac{4a^2}{3m^4}(k_0 a)^2 \frac{\omega_m^2 \Omega_{m0}^2}{\omega_{m0}^2}(\alpha_m - 1). \tag{4.1.9}$$

Optical ω_+ and acoustic ω_- branches are connected with in-phase and anti-phase density oscillations of electrons which participate in longitudinal and transversal motion on the tube.

4.2. Transparency Window for Plasma Waves on the Surface of the Nanotube with a Superlattice

The energy of the electron with the effective mass m_* on the surface of a cylindrical nanotube in a magnetic field parallel to its axis is calculated by Kulik taking into account the quantization of radial motion of electrons in a tube of small thickness [74]:

$$\varepsilon_{lk} = \varepsilon_0 (l + \eta)^2 + \frac{\hbar^2 k^2}{2m_*}, \tag{4.2.1}$$

where $\hbar l$ and $\hbar k$ is the projection of the angular momentum and momentum of electron on the tube axis, $\varepsilon_0 = \hbar^2 / 2m_* a^2$ is the rotational quantum, a is the radius of the tube, $\eta = \Phi / \Phi_0$ is the ratio of the magnetic flux Φ through the cross section of the tube to the flux quantum $\Phi_0 = 2\pi c\hbar / e$ [74]. The equation (4.2.1) describes a set of one-dimensional adjoining subbands whose boundaries $\varepsilon_l = \varepsilon_0(l+\eta)^2$ coincide with the quantized energy levels of the circular motion of the electrons on the tube in the magnetic field. The electron density of states has a root singularity at the boundary of the subzone. The simplest way to take into account the superlattice on the tube is to replace the energy of the longitudinal motion of the electron in the formula (4.2.1) to the expression

$$\varepsilon_k = \Delta(1 - \cos kd).$$
(4.2.2)

This expression is borrowed from the theory of the strong coupling of electrons with the lattice and is often used in the theory of superlattices and layered crystals [31]. As a result of such a substitution, the electron spectrum on the tube becomes

$$\varepsilon_{lk} = \varepsilon_0(l+\eta)^2 + \Delta(1 - \cos kd),$$
(4.2.3)

where d is the period of superlattices, 2Δ is the band width in the energy spectrum of the longitudinal motion of the electron. This band corresponds to the values of the wave number k in the first Brillouin zone $-\pi / d \leq k \leq \pi / d$. The spectrum (4.2.3) describes the set of allowed energy region of electon inside the intervals $\varepsilon_l \leq \varepsilon \leq 2\Delta$ separated by gaps. By analogy with traditional superlattice these bands are called minibands. The electron density of states has a root singularity at the miniband boundaries.

In the random phase approximation the damping decrement of plasma waves with angular momentum $\hbar m$ and the wave number q on the tube is equal to [86]

$$\gamma_m(q) = \frac{\operatorname{Im} P_m(q,\omega)}{\dfrac{\partial}{\partial \omega}\operatorname{Re} P_m(q,\omega)}\Bigg|_{\omega=\omega_m(q)}, \qquad (4.2.4)$$

where $P_m(q,\omega)$ is the polarization operator of electron gas, $\omega_m(q)$ is the plasmon spectrum. The polarization operator is equals

$$P_m(q,\omega) = \frac{1}{\pi a L}\sum_{lk}\frac{f\left(\varepsilon_{(l+m)(k+q)}\right) - f\left(\varepsilon_{lk}\right)}{\varepsilon_{(l+m)(k+q)} - \varepsilon_{lk} - \hbar\omega - i0}, \qquad (4.2.5)$$

where $f(\varepsilon)$ is the Fermi function, L is the the length of the tube.

The dispersion equation for the spectrum and damping of plasma waves has the form [58]

$$1 - \upsilon_m(q)P_m(q,\omega) = 0, \qquad (4.2.6)$$

where

$$\upsilon_m(q) = 4\pi e^2 a \mathrm{I}_m(qa)\mathrm{K}_m(qa) \qquad (4.2.7)$$

is the cylindrical harmonic of electron Coulomb potential on the tube [58], I_m and K_m is the modified Bessel functions, e is the electron charge.

If the electron gas is degenerate, the integration with respect k in the formula (4.2.5) is limited to an interval $\left[-k_l, k_l\right]$, where

$$k_l = \frac{1}{d}\arccos\left(\frac{\varepsilon_l + \Delta - \mu_0}{\Delta}\right) \tag{4.2.8}$$

is the maximum wave number of electrons in a partially filled miniband with the number l, μ_0 is the Fermi energy. The resulting integration over the k the l-miniband contribution to the real part of the polarization operator is defined by parameters

$$c_{\pm} = \frac{\hbar(\omega - \Omega_{\pm})}{2\Delta\sin\dfrac{qd}{2}}. \tag{4.2.9}$$

Here

$$\hbar\Omega_{\pm}(l,m) = \varepsilon_0 m\left[2(l+\eta)\pm m\right], \tag{4.2.10}$$

Ω_{+} is the frequency of direct transitions of electrons $l \to l+m$ between the miniband of the spectrum (4.2.3).

If $c_{\pm}^2 < 1$, from the formula (4.2.5) we obtain

Electron Gas on the Surface of a Nanotube

$$\operatorname{Re} P_{lm}(q,\omega) = -\frac{1}{4\pi^2 ad\Delta\sin\frac{qd}{2}}\left\{\frac{1}{\sqrt{1-c_+^2}}\left[\ln\left|\frac{c_+\operatorname{tg}\frac{1}{2}\left(x_l+\frac{qd}{2}\right)-\left(1-\sqrt{1-c_+^2}\right)}{c_+\operatorname{tg}\frac{1}{2}\left(x_l+\frac{qd}{2}\right)-\left(1+\sqrt{1-c_+^2}\right)}\right|\right.\right.$$

$$\left.-\ln\left|\frac{c_+\operatorname{tg}\frac{1}{2}\left(-x_l+\frac{qd}{2}\right)-\left(1-\sqrt{1-c_+^2}\right)}{c_+\operatorname{tg}\frac{1}{2}\left(-x_l+\frac{qd}{2}\right)-\left(1+\sqrt{1-c_+^2}\right)}\right|\right]-$$

$$-\frac{1}{\sqrt{1-c_-^2}}\left[\ln\left|\frac{c_-\operatorname{tg}\frac{1}{2}\left(x_l-\frac{qd}{2}\right)-\left(1-\sqrt{1-c_-^2}\right)}{c_-\operatorname{tg}\frac{1}{2}\left(x_l-\frac{qd}{2}\right)-\left(1+\sqrt{1-c_-^2}\right)}\right|-\right.$$

$$\left.\left.-\ln\left|\frac{c_-\operatorname{tg}\frac{1}{2}\left(-x_l-\frac{qd}{2}\right)-\left(1-\sqrt{1-c_-^2}\right)}{c_-\operatorname{tg}\frac{1}{2}\left(-x_l-\frac{qd}{2}\right)-\left(1+\sqrt{1-c_-^2}\right)}\right|\right]\right\}, \qquad (4.2.11)$$

where $x_l = k_l d$. When the electrons are completely filled the l-miniband, in the formula (4.2.11) $x_l = \pi$.

In case of $c_\pm^2 > 1$ we find

$$\operatorname{Re} P_{lm}(q,\omega) = +\frac{1}{2\pi^2 ad\Delta\sin\frac{qd}{2}}\times$$

$$\times\left\{\frac{1}{\sqrt{c_+^2-1}}\left[\operatorname{arctg}\frac{c_+\operatorname{tg}\left(x_l+\frac{qd}{2}\right)-1}{\sqrt{c_+^2-1}}+\operatorname{arctg}\frac{c_+\operatorname{tg}\left(x_l-\frac{qd}{2}\right)+1}{\sqrt{c_+^2-1}}\right]-\right. \qquad (4.2.12)$$

$$\left.-\frac{1}{\sqrt{c_-^2-1}}\left[\operatorname{arctg}\frac{c_-\operatorname{tg}\left(x_l+\frac{qd}{2}\right)-1}{\sqrt{c_-^2-1}}+\operatorname{arctg}\frac{c_-\operatorname{tg}\left(x_l-\frac{qd}{2}\right)+1}{\sqrt{c_-^2-1}}\right]\right\}.$$

If we restrict our consideration only the intraband transitions ($m = 0$, $\Omega_{\pm} = 0$, $c_+ = c_- = c = \dfrac{\hbar\omega}{2\Delta\sin\dfrac{qd}{2}}$), the formulas (4.2.11) and (4.2.12) is simplified.

Transition in the formulas (4.2.11) and (4.2.12) towards to the nanotube without superlattice is performed according to the rule

$$d \to 0, \ \Delta \to \infty, \ d^2\Delta \to \hbar^2\big/ m_* . \tag{4.2.13}$$

In this case, the spectrum (4.2.3) becomes (4.2.1) and formula (4.2.11) takes the form

$$\mathrm{Re}\,P_{lm}(q,\omega) = -\frac{m_*}{2\pi^2\hbar^2 qa} \times$$

$$\times \left[\ln\left| \frac{\omega - \left(\Omega_+ + q\upsilon_l + \omega_q\right)}{\omega - \left(\Omega_+ - q\upsilon_l + \omega_q\right)} \right| - \ln\left| \frac{\omega - \left(\Omega_- + q\upsilon_l - \omega_q\right)}{\omega - \left(\Omega_- - q\upsilon_l - \omega_q\right)} \right| \right], \tag{4.2.14}$$

where

$$\upsilon_l = \sqrt{\frac{2}{m_*}}\sqrt{\mu_0 - \varepsilon_l}$$

is the maximal velocity of the electrons in the l-th subzone without superlattice, $\omega_q = \hbar q^2\big/ 2m_*$.

In the absence of interband transitions ($m = 0$) from the formula (4.2.5) we obtain the contribution of the l-miniband into the imaginary part of the polarization operator:

$$\mathrm{Im}\,P_{l0}(q,\omega) = -\frac{1}{4\pi ad\Delta\left|\sin\dfrac{qd}{2}\right|\sqrt{1-c^2}}, \tag{4.2.15}$$

where

$$q < 2k_l, \ 0 < \omega < 2\frac{\Delta}{\hbar}\sin\frac{qd}{2}\sin\left(x_l + \frac{qd}{2}\right).$$

If $q > 2k_l$, the function $\mathrm{Im}\,P_{l0}$ is still equal to (4.2.15) within $\omega_- < \omega < \omega_+$, where

$$\omega_{\pm} = 2\frac{\Delta}{\hbar}\sin\frac{qd}{2}\sin\left(\pm x_l + \frac{qd}{2}\right). \qquad (4.2.16)$$

Figure 15 shows graphs of functions (4.2.15) in these cases. The values of the jump in the points $\omega_{\pm}$ are

$$\frac{1}{4\pi ad\Delta\left|\sin\dfrac{qd}{2}\cos\left(\pm x_l + \dfrac{qd}{2}\right)\right|}.$$

Taking into account the interband transitions ($m \neq 0$) instead of formula (4.2.15) in the vicinity of the frequencies $\Omega_{\pm}$, we obtaine

$$\mathrm{Im}\,P_{lm}^{\pm}(q,\omega) = -\frac{1}{4\pi ad\Delta\left|\sin\dfrac{qd}{2}\right|\sqrt{1-c_{\pm}^2}}, \qquad (4.2.17)$$

as in the formulas (4.2.16) the terms $\Omega_{\pm}$ appear.

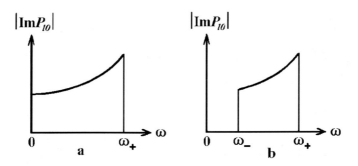

Figure 15. The dependence of the imaginary part of polarization operator (4.2.15) as the functions of the frequency at $q < 2k_l$ (a), $q > 2k_l$ (b).

In the absence of spatial dispersion from the formula (4.2.5) at any temperature we obtain

$$\operatorname{Im} P_m(\omega) = \frac{\pi}{\hbar} \sum_l n_l \left[\delta(\omega - \Omega_-) - \delta(\omega - \Omega_+) \right], \qquad (4.2.18)$$

where n_l is the surface density of electrons in the l-th miniband. In this case, the imaginary part of the polarization operator has sharp jump at frequencies of direct electron transitions between minibands.

From formula (4.2.4) it is seen that for a obtaining of transparency windows for plasma waves in a degenerate electron gas on the surface of the nanotubes, it is sufficient to consider the regions on a plane $q - \omega$ in which $\operatorname{Im} P = 0$. The same regions can be found from the laws of conservation of angular momentum projection and electron momentum projection on the axis of the tube and from the law of conservation of energy under the electron-plasmon absorption. These conservation laws are:

$$\varepsilon_{lk} + \hbar\omega - \varepsilon_{(l+m)(k+q)} = 0. \qquad (4.2.19)$$

Electron Gas on the Surface of a Nanotube 99

From formula (4.2.5) we have seen that the left-hand side of equation (4.2.19) is the argument of the delta function, including into the imaginary part of the polarization operator. In addition, when an quantum electron transfer $(l,k) \to (l+m,k+q)$ is occurs involving the absorption of a plasmon, the Pauli principle must be performed: $\varepsilon_{lk} < \mu_0 < \varepsilon_{(l+m)(k+q)}$. Consequently, after the substitution $k = \pm k_l$ in the equation (4.2.19), we obtain the boundaries of collisionless damping of plasma waves on the tube with a superlattice in a magnetic field:

$$\omega_{\pm}(q) = \Omega_{\pm} + 2\frac{\Delta}{\hbar}\sin\frac{qd}{2}\sin\left(\pm x_l + \frac{qd}{2}\right). \qquad (4.2.20)$$

These equations contains the value $x_l = k_l d$ which determines the position of the Fermi level μ_0 in the l-th miniband. From the formula (4.2.8) it follows that under $x_l = 0$ Fermi energy is located at the "bottom" of the miniband ($\mu_0 = \varepsilon_l$). With the growth of the x_l Fermi energy and the electron density increases. When $x_l = \pi/2$ the level μ_0 is located in the center of the miniband ($\mu_0 = \varepsilon_l + \Delta$) and reaches her "ceiling" on the boundary of the Brillouin zone ($x_l = \pi$, $\mu_0 = \varepsilon_l + 2\Delta$). Thus, when $0 < x_l < \pi/2$ the Fermi level is located in the lower half of the miniband, while $\pi/2 < x_l < \pi$ is at the top.

When $q \to 0$, the difference $\omega_+ - \omega_-$ in the vicinity of each frequency $\Omega_{\pm}$ decreases, the Landau damping region is narrowed in accordance with the behavior of the polarization operator (4.2.18) in the absence of spatial dispersion. This narrowing is occurs at the boundary of the Brillouin zone where the second term on the right-hand side of (4.2.20) is equal to

$$\omega_{\pm} - \Omega_{\pm} = 2\frac{\Delta}{\hbar}\cos x_l . \tag{4.2.21}$$

The shape and dimensions of the regions of Landau damping in the vicinity of the frequencies $\Omega_{\pm}$ are determined by the position of the Fermi level in the miniband. When μ_0 increases from the "bottom" of the l-th miniband to its "ceiling" the expression (4.2.21) in the vicinity of the frequencies $\Omega_{\pm}$ decreases from $2\Delta/\hbar$ to $-2\Delta/\hbar$.

Figure 16 schematically shows Landau damping region between the curves (4.2.20) for various locations of the Fermi energy in the miniband in the vicinity of the frequency Ω_{+} . Outside these regions until the curves

$$\omega_{min} = \Omega_{+} - 2\frac{\Delta}{\hbar}\sin\frac{qd}{2}, \quad \omega_{max} = \Omega_{+} + 2\frac{\Delta}{\hbar}\sin\frac{qd}{2} \tag{4.2.22}$$

located transparency window for plasma waves. The curves of (4.2.22) is the solution of the equation $|c_{+}| = 1$. When $\hbar\Omega_{+} < 2\Delta$, the graph of the curve $\omega_{min}(q)$ (4.2.22) intersects the axis q at the point

$$q_0 = \frac{2}{d}\arcsin\frac{\hbar\Omega_{+}}{2\Delta} .$$

This point tends to the boundary of the Brillouin zone when $\hbar\Omega_{+} \to 2\Delta$. If $\hbar\Omega_{+}$ greater than the width of the miniband 2Δ there is no intersection, i.e., $\omega_{-}(q) > 0$ located in the Brillouin zone.

The areas of Landau damping in the vicinity of the frequency Ω_- are similar to those shown in Figure 16. Note that when $\eta > 1/2$ the boundaries of the miniband are satisfy of inequalities $\varepsilon_{-1} < \varepsilon_0 < \varepsilon_{-2} < ...$ In this case, the frequency of the direct transition of electrons $-1 \to -2$ with $m = -1$ is equal $\Omega_+ = \varepsilon_0(3-2\eta)/\hbar$. In the vicinity of this frequency there exists a branch of the plasmon spectrum with negative helicity.

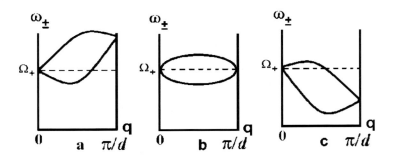

Figure 16. Areas of Landau damping between the curves (4.2.20) at $\cos x_l > 0$ (a), $\cos x_l = 0$ (b) and $\cos x_l < 0$ (c).

In the absence of interband transitions ($\Omega_\pm = 0$) Landau damping regions are bounded by the curve (4.2.20) and the axis q.

In formulas (4.2.20) perform the limit (4.2.13) towards to the nanotube without superlattice. We take into account the Eq. (4.2.8) and

$$\sin x_l = \frac{1}{\Delta}\left[(\mu_0 - \varepsilon_l)(\varepsilon_l + 2\Delta - \mu_0)\right]^{1/2}.$$

Since $qd \ll 1$, from the formulas (4.2.20), taking into account the terms of the order q^2, we find

$$\omega_{\pm} = \Omega_{\pm} \pm \frac{1}{\hbar} qd \left[(\mu_0 - \varepsilon_l)(\varepsilon_l + 2\Delta - \mu_0) \right]^{1/2} + \frac{1}{2\hbar} q^2 d^2 (\varepsilon_l + \Delta - \mu_0)$$

Passing here to the limit (4.2.13), we obtain a parabola

$$\omega_{\pm} = \Omega_{\pm} \pm q\upsilon_l \pm \omega_q,$$

appearing in the Eq. (4.2.14). The maximum speed of the electrons υ_l in the l-th miniband plays the role of the Fermi velocity υ_F of electrons in three-dimensional and two-dimensional electron gas.

The condition of resonant absorption of plasma waves on the tube with a superlattice when $m = 0$ has the form

$$\omega = 2 \frac{\Delta}{\hbar} \sin \frac{qd}{2} \sin \left(k_l + \frac{q}{2} \right) d. \qquad (4.2.23)$$

In the extreme case $qd \ll 1$, $q \ll k_l$, $k_l d < 1$, it takes the usual form in the theory of waves: $\omega/q = \upsilon_l$ is the phase velocity of the wave propagating along the tube is equal to the longitudinal velocity of the electrons.

4.3. Spin Waves on the Surface of a Nanotube with a Superlattice in a Magnetic Field

The electron energy on the surface of a semiconductor nanotube with a superlattice in the magnetic field is [87]

$$\varepsilon_{lk}^{\sigma} = \varepsilon_0 (l + \eta)^2 + \Delta (1 - \cos kd) + g n_{-\sigma} + \sigma \mu B, \qquad (4.3.1)$$

where l and k are the projections of the angular momentum and momentum of the electron on the tube axis, respectively; $\sigma = \pm 1$ is the spin quantum number; $\varepsilon_0 = \left(2 m_* a^2\right)^{-1}$ is the rotational quantum; m_* is the effective electron mass; $\eta = \Phi/\Phi_0$, $\Phi = \pi a^2 B$ is the magnetic flux of the field B through the cross section of the tube; $\Phi_0 = 2\pi c/e$ is the flux quantum; Δ and d are amplitude and period of the modulating potential of the longitudinal superlattice, respectively; g is the Hartree-Fock electron-electron interaction constant [98]; n_σ is the surface density of electrons with the spin projection σ and μ is the spin magnetic moment of electron. Hereinafter, the Planck's constant is set to unity. The first term in Eq. (4.3.1) refers to the quantized levels of the circular motion of electrons on the tube in the magnetic field, the second term is the energy of the longitudinal motion of the electrons, and the third and fourth terms are the exchange shift and the spin splitting of the levels, respectively. The energy spectrum of the longitudinal motion of electrons consists of narrow minibands with the widths 2Δ separated by energy gaps. The minibands can overlap. Small-radius tubes correspond to the case with a small number of occupied lower minibands.

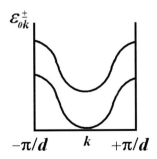

Figure 17. Electron energy (4.3.1) in two overlapping minibands $0^\pm$.

Figure 17 shows spectrum (4.3.1) in the first Brillouin zone $-\pi/d < q < \pi/d$ when two lower spin-split minibands $0^{\pm}$ ($l=0, \sigma=\pm1$) overlap. We consider the case of $\eta < 1/2$ when the positions of the lower boundaries $\varepsilon_l^{\pm}$ of the minibands satisfy the inequalities $\varepsilon_0^- < \varepsilon_0^+ < \varepsilon_{-1}^- < \varepsilon_{-1}^+ < ...$ The miniband overlapping region $\left[\varepsilon_0^+, \varepsilon_0^- + 2\Delta \right]$ in Figure 17 has the width $2\Delta - \Omega$ with $\Omega = g\delta n + 2\mu B$, $\delta n = n_- - n_+$.

In the random-phase approximation, the dispersion relation for the spectrum of transverse spin waves on a tube with a superlattice in the magnetic field has the form [58]

$$1 - \frac{g}{2\mu^2} \chi_{\pm} (m,q,\omega) = 0, \tag{4.3.2}$$

where $\chi_{\pm} = \chi_{xx} \pm i\chi_{yx}$ are the circular components of the dynamic spin susceptibility tensor of the electron gas with spectrum (4.3.1). These components have the form

$$\chi_{\pm} (m,q,\omega) = \frac{2\mu^2}{S} \sum_{lk} \frac{f\left(\varepsilon_{(l+m)(k+q)}^{\mp}\right) - f\left(\varepsilon_{lk}^{\pm}\right)}{\omega + \varepsilon_{lk}^{\pm} - \varepsilon_{(l+m)(k+q)}^{\mp} + i0}, \tag{4.3.3}$$

where m, q and ω are the magnon angular momentum, momentum and energy, respectively; f is the Fermi distribution function; and S is the surface of the tube. The plus (minus) sign in $\chi_{\pm}$ corresponds to transverse Landau-Silin spin waves [99-102] with a negative (positive) chirality. These waves in bulk conductors were predicted by Landau [99] and Silin [100]. Their properties were considered in [101-103].

The solution of Eq. (4.3.2) for a degenerate electron gas depends on the position of the Fermi level μ_0. If the electron density n satisfies the inequality

$$n < \frac{1}{2\pi^2 a}\left(k_0^- + k_0^+\right), \qquad (4.3.4)$$

the Fermi level occurs in the miniband overlapping region in Figure 17. Inequality (4.3.4) involves

$$k_0^\pm = \frac{1}{d}\arccos\frac{\varepsilon_0^\pm + \Delta - \mu_0}{\Delta},$$

which is the maximum electron momentum in the miniband $0^\pm$. If the minibands do not overlap and the level μ_0 is situated in the second miniband then k_0^- in Eq. (4.3.4) should be replaced by π/d.

The graphical analysis of Eq. (4.3.2) in the case $\varepsilon_0^+ < \mu_0 < \varepsilon_0^- + 2\Delta$ indicates that each m value i. e. each spin-flip $- \to +$ electron transition $0^- \to m^+$ between the minibands $l=0$ and $l=m$ corresponds to two branches of the magnon spectrum with a positive chirality. These branches are situated between the frequencies of single-electron transitions between the minibands

$$\Omega_\pm = \varepsilon_0 m\left[2\eta \pm m\right] + \Omega.$$

In the long-wavelength limit ($2\Delta\left|\sin\frac{qd}{2}\right| \ll |\omega - \Omega_\pm|$), the magnon spectrum with a positive chirality reads

$$\omega_\pm(q) = \omega_\pm^0 + \alpha_\pm \sin^2 \frac{qd}{2},$$ (4.3.5)

where

$$\omega_\pm^0 = \frac{1}{2}\Big[\Omega_+ + \Omega_- - \upsilon\big(k_0^- - k_0^+\big)\Big] \pm$$
$$\pm \frac{1}{2}\Big[(\Omega_+ - \Omega_-)^2 - 2\upsilon\big(k_0^- - k_0^+\big)(\Omega_+ + \Omega_-) +$$ (4.3.6)
$$+ \upsilon^2\big(k_0^- - k_0^+\big) - 4\upsilon\big(k_0^+\Omega_+ - k_0^-\Omega_-\big)\Big]^{1/2}$$

are the limiting frequencies of the modes,

$$\upsilon = \frac{g}{2\pi^2 a},$$ (4.3.7)

$$\alpha_\pm = 2\Delta \frac{\sin k_0^+ d\left(\omega_\pm^0 - \Omega_+\right)^2 + \sin k_0^- d\left(\omega_\pm^0 - \Omega_-\right)^2}{k_0^+ d\left(\omega_\pm^0 - \Omega_+\right)^2 - k_0^- d\left(\omega_\pm^0 - \Omega_-\right)^2}.$$

If the minibands 0^- and 0^+ do not overlap, k_0^- in Eqs. (4.3.6) and (4.3.7) must be replaced by π/d. The spectrum of negative-chirality spin waves can be found from Eqs. (4.3.6) and (4.3.7) by exchanging the spin indices $- \leftrightarrow +$ and a sign change of Ω.

In the case of weak electron-electron interaction $\upsilon \ll d\Omega_\pm$, we find from Eqs. (4.3.6) and (4.3.7)

$$\omega_\pm^0 = \Omega_\pm - \upsilon k_0^\mp,$$ (4.3.8)

$$\alpha_\pm = \mp 2\Delta \frac{\sin k_0^\mp d}{k_0^\mp d},\tag{4.3.9}$$

where $\upsilon k_0^\mp$ is the depolarization frequency shift. The upper (lower) branch ω_+ (ω_-) of the magnon spectrum has a negative (positive) chirality.

Collisionless damping of spin waves is given by the imaginary part of susceptibility (4.3.3). In the case of a degenerate electron gas it is

$$\operatorname{Im}\chi_- = \frac{\mu^2}{2\pi a d}\sum_l \left[4\Delta^2\sin^2\frac{qd}{2} - (\omega - \Omega_+)^2\right]^{-\frac{1}{2}},\tag{4.3.10}$$

where

$$\Omega_+ - 2\Delta\sin\frac{qd}{2} < \omega < \Omega_+ + 2\Delta\sin\frac{qd}{2};$$

$$\operatorname{Im}\chi_- = -\frac{\mu^2}{2\pi a d}\sum_l \left[4\Delta^2\sin^2\frac{qd}{2} - (\omega - \Omega_-)^2\right]^{-\frac{1}{2}},\tag{4.3.11}$$

where $\Omega_- - 2\Delta\sin\frac{qd}{2} < \omega < \Omega_- + 2\Delta\sin\frac{qd}{2}$.

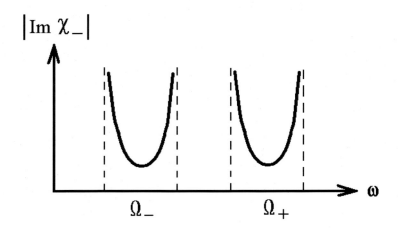

Figure 18. Frequency dependence of the imaginary part (4.3.10) and (4.3.11) of the susceptibility.

The frequency dependence of imaginary part (4.3.10) and (4.3.11) of susceptibility is schematically shown in Figure 18 for the case of

$$\Omega_- > 0, \; \varepsilon_0^+ < \mu_0 < \varepsilon_0^- + 2\Delta.$$

The Landau damping of spin waves is nonzero in the Stoner sectors of the $q-\omega$ plane bounded by the curves

$$\omega_\pm = \Omega_\pm \pm 2\Delta \sin\frac{qd}{2}.$$

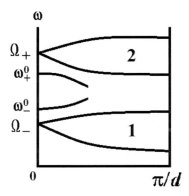

Figure 19. Magnon spectrum (4.3.5) and regions 1 and 2 of the Landau damping of spin waves on the tube.

Dispersion curves (4.3.5) are situated outside the Stoner sectors, i. e. the spin waves considered in this work are undamped. Figure 19 shows dispersion curves (4.3.5) in regions 1 and 2 of the Landau damping. To observe the effects associated with these modes the distances between the edges $\upsilon k_0^\pm = 2\pi^2 a \upsilon n_\pm$ of the Stoner sector and the limiting frequencies must exceed both the thermal and impurity spread of the electron energy levels.

Conclusion

In the framework of the hydrodynamic approach, the plasma waves on the surface of a nanotube with a longitudinal superlattice are considered in the Subsection 4.1. Not only longitudinal electron current but also transversal one has been taken into consideration. It has been shown that both optical and acoustical plasmons could propagate along the tube with one sort of carrier.

Within the framework of the model electron energy spectrum on the nanotube surface with a superlattice in a magnetic field, an exact expression for the polarization operator of a degenerate electron gas in the

Subsection 4.2 was obtained. The shape and size of the plasma waves Landau damping regions on the tube throughout the Brillouin zone were calculated. The influence on these areas of the position of the Fermi level in the miniband was considered. The conditions for the resonance absorption of plasmon on the tube by electrons were found. The limiting transition towards the nanotube without superlattice was performed.

Electron spin waves on the surface of a semiconductor nanotube with a superlattice in a magnetic field have been considered in the Subsection 4.3. The spin-wave spectra and regions of collisionless wave damping have been found. It has been shown that the spin waves do not exhibit damping on small-radius tubes with a degenerate electron gas.

GENERAL CONCLUSION

The Laplace transform is a convenient method for calculating the thermodynamic function of electron gas on the semiconductor nanotube surface in a longitudinal magnetic field. In the effective mass approximation this method allows obtaining exact expressions for the density of states and thermodynamic quantities of degenerate and nondegenerate electron gas. It has been demonstrated here that the thermodynamic quantities of degenerate electron gas experience oscillations similar to the de Haas-van Alphen and the Aharonov-Bohm oscillations. The de Haas-van Alphen oscillations are caused by the passage of Fermi energy through the peaks of the density of states at subband edges. The period of these oscillations is determined by the electron effective mass and the semiconductor tube radius. They are retained in the absence of a magnetic field. To observe these oscillations, one would need to change the Fermi energy of electrons, similar as it is done for the two-dimensional electron gas. The Aharonov-Bohm oscillations remain only in the nondegenerate electron gas. These oscillations are caused by the change of the magnetic flux within the tube section. The period of these oscillations equals to the magnetic flux quantum. The monotonic parts of thermodynamic quantities coincide with the quantities for the two-dimensional electron gas that are obtained by cutting the tube along its generatrix and transforming it into a strip. The dependence of the nondegenerate electron gas rotational heat capacity against the temperature in the absence of a magnetic field shows a peak at a certain temperature. For typical semiconductor nanotubes (e.g., GaAs) this temperature is within a premelting temperature range.

Superlattice at the surface of a nanotube has a significant impact on its properties. It can be obtained by embedding fullerenes or other additives to the nanotube or when the nanotube is attached to a substrate for charge exchange. In the absence of a superlattice, the tube spectrum in a longitudinal magnetic field is a collection of one-dimensional subbands located next to each other and having nonequidistant boundaries. Periodic modulating potential artificially created at the surface of the tube converts the spectrum into a system of minibands, the widths of which are determined by the amplitude of the modulating potential. In a longitudinal magnetic field its amplitude and period depend on the magnetic field strength. Energy gaps separating the minibands have widths defined by the ratio of the miniband width to the magnitude of the rotational quantum and depend on magnetic field strength. Density of electronic states has a root singularity at miniband borders. As the radius of the tube increases, the minibands overlap resulting in a continuous spectrum. Here the state density, chemical potential, energy and heat capacity of a degenerate and non-degenerate electron gas at the surface of a nanotube with metallic conductivity character in a longitudinal magnetic field have been given. We show that abovementioned thermodynamic values include monotonic and oscillating components. In agreement with Pauli principle monotonic heat capacity component of a degenerate electron gas is proportional to the temperature. Heat capacity displays de Haas-van Alphen type oscillations due to the passage of state density root singularity through the Fermi boundary with a change in electron density. These oscillations persist in the absence of a magnetic field. Heat capacity also displays Aharonov-Bohm type oscillations when magnetic field flux through cross-sectional of the tube is varied. Heat capacity studies allow observation of the transition of modulating potential from the localized gaps mode to the mode of free motion along the tube.

Simplified conductivity models are usually used for studying electromagnetic waves propagation in the cylindrical geometry systems, for example, in nanotubes. The metal cylinder conductivity is often believed to be endless, and the dielectric permittivity of the matter in which cylinder is dipped is considered to be constant or only frequency dependent. Conductivity's spatial dispersion is usually not taken into account. Nonetheless, the electromagnetic field's nature in the tube, its waveguide characteristics are sensitive to the surface currents. Therefore, the electron gas conductivity tensor components calculation problem with allowance for the spatial and time dispersion is worth consideration. Given in this chapter conductivity tensor components may be used in obtaining a dispersion equation for electromagnetic waves in the tube. The example of longitudinal conductivity showed what data of electron gas is possible to obtain by measuring conductivity. Particularly, imaginary part of conductivity experiences de Haas - van Alphen and Aharonov - Bohm oscillations types with the electron density and magnetic field changes. The oscillation periods measurement enables one to identify the electron effective mass and the combination of universal constants included into flux quantum.

The obtained in the present chapter formulas for the conductivity tensor components may be applied for studying electromagnetic wave propagation in nanotubes with superlattices based on $Al_x Ga_{1-x} As / Ga As$, $InGa As / Ga As$, $In As / Ga As$, $GeSi / Si$ heterojunctions and in carbon nanotubes in the regime of metallic conductivity. The real part of conductivity determines the wave energy absorbed by electrons. In the degenerated electron gas, this is nonzero in the areas of Landau collisionless damping. Knowing the positions of transparency windows for the waves, it is possible to improve the waveguide characteristics of nanotubes. The imaginary part of conductivity is included into the dispersion equation for electromagnetic wave spectrum. This has the resonance singularities at frequencies of electron direct transitions between minibands. Usually, near these frequencies there exist new branches in the wave spectrum and related band transparency. Observation of conductivity oscillations of de Haas-van Alphen type allows determining the electron effective mass, Fermi momentum, rotational quantum and superlattice parameters d and Δ. These values are included in the oscillation amplitude and period expressions. Revealing the instant of appearing the beats under variation of the nanotube parameters gives the opportunity to obtain the ratio of Fermi energy to miniband width.

The expressions obtained in this chapter for the components of the conductivity tensor can be used to study the propagation of electromagnetic waves in a tube. The real part of the conductivity differs from zero in the regions of the Landau collisionless damping. The transparency windows for waves are located outside these areas. Knowledge of the positions of these windows allows us to improve the waveguide characteristics of the tube. The imaginary part of the conductivity determines the spectrum of the waves. It has resonant peculiarities at the frequencies of electron transitions between minibands. Near these frequencies there should be new branches of the spectrum of the waves and new windows of transparency for these waves. The appearance in the components of the conductivity tensor of such parameters of the electronic spectrum as the electron effective mass, rotational quantum, Fermi energy, period and amplitude of the modulating potential allows these parameters to be obtained in experiments on measuring the high frequency conductivity of nanotubes. Suitable objects for these studies are rolled-up hetero systems $Al_xGa_{1-x}As/GaAs$, $InGaAs/GaAs$, $InAs/GaAs$ as well as graphene-based semiconductor nanostructures.

In the framework of the hydrodynamic approach, the plasma waves on the surface of a nanotube with a longitudinal superlattice are considered. Not only longitudinal electron current but also transversal one has been taken into consideration. It has been shown that both optical and acoustical plasmons could propagate along the tube with one sort of carrier. Study of propagation of plasma waves along the tube are very topical problem because it allows to determine the waveguide characteristics of the tube. In addition, it is possible to obtain additional information about the dynamic characteristics of the conduction electrons on the curved surface. The presence of an additional parameter – the curvature of the structure – enriches the picture of wave phenomena by increasing the number of ways to control the properties of the system. In particular, the rotational quantum contains the characteristic of circular motion of electrons on the tube – transverse effective mass. It may differ from the longitudinal mass. Superlattice adds new features to the picture of wave propagation. It is associated with additional parameters – the period and amplitude of the modulating potential. Characteristics of the tube – form and sizes of windows transparency of waves and their spectrum and damping – are sensitive to these parameters. This allows them to be determined by analyzing the properties of the waves.

The chapter used a simple model spectrum of electrons simulating the superlattice on the tube. This allowed within the model adopted in the random phase approximation to obtain an exact expression for the polarization operator of the electron gas. As a result, the shape and size of windows transparency for plasma waves were determined in the entire Brillouin zone. They were obtained by analysis of the imaginary part of the polarization operator and with the help of conservation laws. The results can be used in the study of plasma waves in semiconductor superlattices on a base of $Al_x Ga_{1-x} As/Ga As$, $InGa As/Ga As$, $In As/Ga As$, $GeSi/Si$ and in carbon nanotubes in a metal conduction mode.

The results of this chapter can be used in studying the magnetic scattering of neutrons by the spin magnetization current of conduction band electrons on a tube. The cross sections of scattering by spin waves and Stoner excitations are of interest. This problem was solved earlier for a two-dimensional electron gas on a plane. The curvature of a cylinder should manifest itself in additional features of the scattering cross section. The electron-electron interaction constant, the amplitude and period of the modulating potential can be found by measuring the depolarization frequency shift and group velocity of spin waves on a tube.

ACKNOWLEDGMENTS

The authors are thankful to T. I. Rashba for help during the manuscript preparation.

REFERENCES

[1] Iijima, S. *Nature* 1991, *vol* **354**, 56-58.

[2] Dresselhaus, M. S.; Dresselhaus, G.; Eklund, P. C. *Science of Fullerens and Carbon Nanotubes*; Acad. Press: New York, USA, 1996; pp. 1-919.

[3] Saito, R.; Dresselhaus, G.; Dresselhaus, M. S. *Physical properties of carbon nanotubes*; Imperial College Press: London, UK, 1998; pp. 1-259.

[4] Magarill, L. I.; Chaplik, A. V.; Entin, M. V. *Usp. Fiz. Nauk* 2005, *vol* **175**, 995-1000.

[5] Mahan, G. D. *Phys. Rev.* 2004, *vol* **B69**, 125407 (7).

[6] Ermolaev, A. M.; Rashba, G. I. *J. Phys.: Condens. Matter* 2008, vol **20**, 175212 (7).

[7] Rumer, J. B. *JETP* 1948, *vol* **18**, 1081-1095.

[8] Rumer, J. B.; Ryvkin M. Sh. *Thermodynamics, statistical physics and kinetics*; Nauka: Moscow, USSR, 1977; pp. 1-552.

[9] Wendler, L.; Pechstedt, R. *Phys. Status Solidi* 1986, *vol* **B138**, 197-217.

[10] Wendler, L.; Kandler, E. *Phys. Status Solidi* 1993, *vol* **B 177**, 9-67.

[11] Chun, I. S.; Verma, V. B.; Elarde, V. C.; Kim, S. W.; Zuo, J. M.; Coleman, J. J.; Li, X. *Journ. of Cryst. Growth* 2008, *vol* **310**, 2353-2358.

[12] Smith, B. W.; Benes, Z.; Luzzi, D. E.; Fischer, J. E.; Walters, D. A.; Casavant, M. J.; Schmidt, J.; Smalley, R. E. *Appl. Phys. Let.* 2000, *vol* **77**, 663-665.

[13] Walters, D. A.; Casavant, M. J.; Qin, X. C.; Huffman, C. B.; Boul, P. J.; Ericson, L. M.; Haroz, E. H.; O'Connell, M. J.; Smith, K.; Colbert, D. T.; Smalley, R.E. *Chem. Phys. Lett.* 2001, *vol* **338**, 14-20.

[14] Sun, K. J.; Wincheski, R. A.; Park, C. *Journ. of Appl. Phys.* 2008, *vol* **103**, 023908 (6).

[15] da Kosta, R. C. T. *Phys. Rev.* 1981, *vol* **A23**, 1982-1987.

[16] Magarill, L. I.; Romanov, D. A.; Chaplik, A. V. *JETP* 1998, *vol* **113**, 1411- 1428.

[17] Geyler, V. A.; Margulis, V. A.; Shorohov, A. B. *JETP* 1999, *vol* **115**, 1450-1462.

[18] Ando, T. *J. Phys. Soc. Jpn.* 2005, *vol* **74**, 777- 817.

[19] Tsuji, N.; Takajo, S.; Aoki, H. *Phys. Rev.* 2007, *vol* **B75**, 153406 (4).

[20] Sasaki, K. *Phys. Rev.* 2002, *vol* **B 65**, 195412 (12).

[21] Vitlina, R. Z.; Magarill, L. I.; Chaplik, A. V. *Pis'ma v JETP* 2007, *vol* **86**, 132- 134.

[22] Wu-Sheng Dai, Xie, M. *Phys. Rev.* 2004, *vol* **E70**, 016103 (15).

[23] Kuznetsov, V. L.; Mazov, I. N.; Delidovich, A. I.; Obraztsova, E. D.; Loiseau, *Phys. Status Sol.* 2007, *vol* **B244**, 4165-4169.

[24] Cordeiro, C. E.; Delfino, A.; Frederico, T. *Phys. Rev.* 2009, *vol* B**79**, 035417 (11).

[25] Cordeiro, C. E.; Delfino, A.; Frederico, T. *Carbon* 2009, *vol* **47**, 690-695.

[26] Sondheimer, E. H.; Wilson, A. H. *Proc. Roy. Soc. (London)* 1951, *vol* **A210**, 173 -190.

[27] Kubo, R. *Statistical Mechanics*; North-Holland Publ. Company: Amsterdam, Holland, 1965; pp. 1-434.

[28] Lifshitz, I. M.; Azbel, M. Ya.; Kaganov, M. I. *The Electron Theory of Metals*; Nauka: Moscow, USSR, 1971; pp. 1-416.

[29] Lifshitz, I. M.; Peschanski, V. G. *Zh. Eksp. Teor. Fiz.* 1958, *vol* **35**, 1251- 1264.

[30] Lifshitz, I. M.; Peschanski, V. G. *Zh. Eksp. Teor. Fiz.* 1960, *vol* **38**, 188-193.

[31] Bass, F. G.; Bulgakov, A. A.; Tetervov, A. P. *Quality Properties of Conductors with Superlattices*; Nauka: Moscow, USSR, 1989; pp. 1-288.

[32] Herman, M. A. *Semiconductor Superlattices;* Akademie-Verlag: Berlin, Germany, 1986; pp. 1-240.

[33] Volosknikova, O. P.; Zavyalov, D. V.; Kryuchkov, S. V. *Proceeding of the International Meeting titled "Radiation Physics of Solids"*: Sevastopol, Ukraine, 2007; pp. 645-649.

[34] Pfeiffer, C. A.; Economou, E. N.; Ngai, K. L. *Phys. Rev.* 1974, *vol* **B10**, 3038-3051.

[35] Vasconcelos, E. F.; Oliveira, N. T.; Farias, G. A. *Phys. Rev.* 1991, *vol* **B44**, 12621-12623.

[36] Almeida, N. S.; Farias, G. A.; Oliveira, N. T.; Vasconcelos, E. F. *Phys. Rev.* 1993, *vol* **B48**, 9839-9845.

[37] Lin, M. F.; Shung, K. W. K. *Phys. Rev.* 1993, *vol* **B47**, 6617-6624.

[38] Sato, O.; Tanaka, Y.; Kobayashi, M.; Hasegawa, A. *Phys. Rev.* 1993, *vol* **B48**, 1947-1950.

[39] Lin, M. F.; Shung, K. W. K. *Phys. Rev.* 1993, *vol* **B48**, 5567-5571.

[40] Longe, P.; Bose, S. M. *Phys. Rev.* 1993, *vol* **B48**, 18239-18243.

[41] Kushwaha, M. S.; Djafari-Rouhani, B. *Phys. Rev.* 2003, *vol* **B67**, 245320 (19).

[42] Kushwaha, M. S.; Djafari-Rouhani, B. *Phys. Rev.* 2005, *vol* **B71**, 195317 (21).

[43] Moradi, A.; Khosravi, H. *Phys. Lett.* 2007, *vol* **A371**, 1-6.

[44] Moradi, A.; Khosravi, H. *Phys. Rev.* 2007, *vol* **B76**, 113411 (11).

[45] Prinz, V. Ya.; Chehovskiy, A. V.; Preobrazhenskii, V. V.; Semyagin, B. R.; Gutakovsky, A. K. *Nanotechnology* 2002, *vol* **13**, 231-233.

[46] Chun, I.; Verma V. Journ. of Cryst. Growth. 2008, *vol* **310,** 235-238.

[47] Keldysh, L. V. *Fiz. Tverd. Tela* 1962, *vol* **4**, 2265-2267.

[48] Esaki, L.; Tsu, R. *IBM J. Dev.* 1970, *vol* **14**, 61-65.

[49] Fetter, A. *Ann. Phys.* 1974, *vol* **88,** 1-25.

[50] Sarma, D.; Quinn, J. J. *Phys. Rev.* 1982, *vol* B**25**, 7603-7618.

[51] Tselis, A. C.; Quinn, J. J. *Phys. Rev.* 1984, *vol* B**29**, 2021-2027.

[52] Tselis, A. C.; Quinn, J. J. *Phys. Rev.* 1984, *vol* B**29**, 3318-3335.

118   *A. M. Ermolaev and G. I. Rashba*

[53] Wei-ming, Q.; Kirczenow, G. *Phys. Rev.* 1987, *vol* B**36**, 6596-6601.

[54] Golden, K.; Kalman, G. *Phys. Rev.* 1995, *vol* **B 52**, 14719-14727.

[55] Dragunov, V.; Neizvestnyi, I.; Gridchin, V. *Fundamentals of Nanoelectronics*; Logos: Moscow, Russia, 2006; pp. 1-496.

[56] Yannouleas, C.; Bogachek, E.; Landman, U. *Phys. Rev.* 1996, *vol* B **53**, 10225-10236.

[57] Bogachek, E.; Gogadze, G. *JETP* 1975, *vol* **40**, 308-310.

[58] Ermolaev, A. M.; Rashba, G. I. *Handbook of Functional Nanomaterials*. Vol. 4 – *Properties and Commercialization*. Chapter 11. *Collective Excitations of Electron Gas on the Nanotube Surface in a Magnetic Field: Magnetoplasma and Spin Waves, Zero Sound*; Nova Science Publishers: New-York, USA, 2013, pp. 213-246.

[59] Ermolaev, A. M.; Solyanik, M. A. *Bulletin KhNU named after V. N. Karazin, Ser. "Physics"* 2008, *vol* **821**, 9-13.

[60] Ermolaev, A. M.; Rashba, G. I.; Solyanik, M. A. *Europ. Phys. Journ.* 2010, *vol* **B73**, 383-388.

[61] Ermolaev, A. M.; Rashba, G. I.; Solyanik, M. A. *Low Temperature Physics* 2011, *vol* **37**, 824-828.

[62] Ermolaev, A. M.; Kofanov, S. V.; Rashba, G. I. *Advances in Condensed Matter Physics*, 2011, *vol* **2011**, 901848 (7).

[63] Ermolaev, A. M.; Rashba, G. I. *Physica B* 2014, *vol* **451**, 20-25.

[64] Ermolaev, A.M.; Rashba, G. I. *Bulletin KhNU named after V. N. Karazin, Ser. "Physics"*, 2013, *vol* **1075**, 14-19.

[65] Ermolaev, A. M.; Rashba, G. I. *Solid State Commun.* 2014, *vol* **192**, 79-81.

[66] Ermolaev, A. M.; Rashba, G. I. *Bulletin KhNU named after V. N. Karazin, Ser. "Physics"* 2014, *vol* **1135**, 10-15.

[67] Ermolaev, A. M.; Rashba, G. I. *Physics of the Solid State* 2014, *vol* **56**, 1696– 1699.

[68] Prudnikov, A. P.; Brychkov, Yu. A.; Marichev, O. I. *Integrals and Series. Elementary functions*; Nauka: Moscow, USSR, 1981; pp. 1-800.

[69] Prudnikov, A. P.; Brychkov, Yu. A.; Marichev, O. I. *Integrals and Series. Special functions*; Nauka: Moscow, USSR, 1983; pp. 1-753.

[70] Bateman, G.; Erdelyi, A. Higher Transcendental Functions. Volume 2; Nauka: Moscow, USSR, 1974; pp. 1-295.

[71] Landau, L. D.; Lifshits, E. M. *Statistical physics. Part 1*; Nauka: Moscow, USSR, 1995; pp. 1-616.

[72] Ermolaev, A. M., Rashba, G. I. *Introduction to statistical physics and thermodynamics*; KhNU named after V. N. Karazin: Kharkiv, Ukraine, 2004; pp. 1-516.

[73] Ando, T.; Fowler, A.; Stern, F. *Rev. Mod. Phys.* 1982, *vol* **54**, 437-667.

[74] Kulik, I. O. *JETP Lett.* 1970, *vol* **11**, 275-278.

[75] Peeters, F. M.; Vasilopoulos, P. *Phys. Rev.* 1992, *vol* **B 46**, 4667-4680.

[76] Yanke, E.; Emde, F.; Lyosh, F. *Special Functions*; Nauka: Moscow, USSR, 1977; pp. 1-343.

[77] Peschanski, V. G. *Fiz. Nizk. Temp.* 1997, *vol* **23**, 47-51.

[78] Erdelyi, A. *Asymptotic Expansions*; State Publishing House of Physical-Mathematical Literature: Moscow, USSR, 1962; pp. 1-128.

[79] Kulik, I. O. *Zh. Eksp. Teor. Fiz.* 1970, *vol* **58**, 2171- 2175.

[80] Zvyagin, A. A.; Krive, I. V. *Fiz. Nizk. Temp.* 1995, *vol* **21**, 687-716.

[81] Gvozdikov, V. M. *Fiz. Nizk. Temp.* 2000, *vol* **26**, 648-657.

[82] Kulik, I. O. *Fiz. Nizk. Temp.* 2010, *vol* **36**, 1057-1065.

[83] Abrikosov, A. A.; Gor'kov, L. P.; Dzyaloshinsky, I. E. *Quantum field theory methods in statistical physics*; Fizmatgiz: Moscow, USSR, 1962; pp. 1-444.

[84] Mahan, G. D. *Many particle physics*; Plenum Press: New York, USA, 1990; pp. 1-1032.

[85] Ermolaev, A. M.; Rashba, G. I.; Solyanik, M. A. *Physica* 2011, *vol* **B406**, 2077-2080.

[86] Ermolaev, A. M.; Rashba, G. I.; Solyanik, M. A. *Low Temp. Phys.* 2012, *vol* **38**, 511-516.

[87] Ermolaev, A. M.; Rashba, G. I.; Solyanik, M. A. *Low Temp. Phys.* 2012, *vol* **38**, 1209-1215.

[88] Fedoryuk, M. V. *Asymptotics: Integrals and Series*; Nauka: Moscow, USSR, 1987; pp. 1-544.

120   *A. M. Ermolaev and G. I. Rashba*

[89] Vedernikov, A. I.; Govorov, A. O.; Chaplik, A. V. *JETP* 2001, *vol* **120**, 979- 985.

[90] Ermolaev, A. M.; Rashba, G. I.; Solyanik, M. A. *Fiz. Tverd. Tela* 2011, *vol* **53**, 1594-1598.

[91] Eminov, P. A.; Perepelkina, Yu. V.; Sezonov, Yu. I. *Phys. Solid State* 2008, *vol* **50**, 2220-2224.

[92] Prinz, V. Ya.; Seleznev, V. A.; Samoylov, V. A.; Gutakovsky, A. K. *Microelectron. Eng.* 1996, *vol* **30**, 439-442.

[93] Prinz, V. Ya. *Physica* 2004, *vol* **E24**, 54-62.

[94] Vitlina, R. Z.; Magarill, L. I.; Chaplik, A. V. *JETP* 2008, *vol* **133**, 906-913.

[95] Lee, J.; Kim, H.; Kahng, S. J.; Kim, G. *Nature* 2002, *vol* **415**, 1005-1008.

[96] Korol, A. N. *Pis'ma Zh. Eksp. Teor. Fiz.* 1994, *vol* **59**, 659-662.

[97] Shin, H.-J.; Clair, S.; Kim, Y.; Kawai, M. *Nature Nanotech.* 2009, *vol* **4**, 567-570.

[98] White, R. M. *Quantum Theory of Magnetism*; Springer-Verlag: Berlin-Heidelberg, Germany, 2007; pp. 1-359.

[99] Landau, L. D. *JETP* 1957, *vol* **32**, 59-66.

[100] Silin, V. P. *JETP* 1958, *vol* **35**, 1243-1250.

[101] Kondrat'ev, A. S.; Kuchma, A. E. *Electron Liquid of Normal Metals*; Leningrad State University: Leningrad, USSR, 1980; pp. 1-198.

[102] Kondrat'ev, A. S.; Kuchma, A. E. *Lectures on the Theory of Quantum Liquids*; Leningrad State University: Leningrad, USSR, 1989; pp. 1-264.

[103] Ermolaev, A. M.; Ulyanov, N. V. *Landau-Silin Spin Waves in Conductors with Impurity States*; Lambert: Saarbrucken, Germany, 2012; pp. 1-109.

BIOGRAPHICAL SKETCH

Alexander M. Ermolaev

Affiliation: Theoretical Physics Department named after academician I. M. Lifshits, V. N. Karazin Kharkiv National University, Kharkov, Ukraine

Education: Kharkiv State University

Business Address: 4 Svobody Sq., Kharkiv, 61022, Ukraine

Research and Professional Experience: Author more than 200 scientific papers and textbooks. He was a supervisor of 5 PhD students. Author of textbooks "Introduction in statistical physics and thermodynamics" (2004, 516 p.), "Lectures on quantum statistics and kinetics"
(2012, 504 p.). The sphere of scientific interests is theoretical physics, in particular the solid state theory. He developed the theory of magnetoimpurity states of electrons in solids. The new branches in the spectrum of collective excitations of solids with magnetoimpurity states of electrons was predicted.

Professional Appointments: Professor of the Theoretical Physics Department name after academician I. M. Lifshits

Honors: Excellence in Education of Ukraine, winner of the Municipal Administration Award named after Academician K. D. Sinel'nikov, awarded a medal named V. N. Karazin

Georgiy I. Rashba

Affiliation: Theoretical Physics Department named after academician I. M. Lifshits, V. N. Karazin Kharkiv National University, Kharkov, Ukraine

Education: Kharkiv State University

Business Address: 4 Svobody Sq., Kharkiv, 61022, Ukraine

Research and Professional Experience: Author more than 70 scientific papers and textbooks. Author of textbooks "Introduction in statistical physics and thermodynamics" (2004, 516 p.), "Lectures on quantum statistics and kinetics" (2012, 504 p.). The sphere of scientific interests is theoretical physics, in particular the solid state theory. He developed the theory of magnetoimpurity states of electrons in nanosystems. The new branches in the spectrum of collective excitations of nanosystems with magnetoimpurity states of electrons was predicted.

Professional Appointments: Associate Professor of the Theoretical Physics Department name after academician I. M. Lifshits

In: Electron Gas: An Overview
Editor: Tata Antonia

ISBN: 978-1-53616-428-2
© 2019 Nova Science Publishers, Inc.

Chapter 2

STUDY OF THE TRANSPORT OF CHARGE CARRIERS IN MATERIALS WITH DEGENERATE ELECTRON GAS

Vilius Palenskis[*] *and Evaras Žitkevičius*
Faculty of Physics, Vilnius University, Vilnius, Lithuania
Department of Electronic Systems,
Vilnius Gediminas Technical University, Vilnius, Lithuania

ABSTRACT

This study is addressed to the stochastic description of the effective density of the randomly moving (RM) electrons in metals and other materials with degenerate electron gas. It is written in the accessible form for researchers, engineers and students without an extensive background of quantum mechanics of solid-state physics. The chapter begins with the interpretation of the basic transport characteristics of the metals, superconductors in the normal state, and very strongly doped semiconductors with degenerate electron gas. An application of the effective density of RM electrons leads one simply to explain the

---

[*] Corresponding Author's E-mail: vilius.palenskis@ff.vu.lt.

conductivity of metals, and the electron transport characteristics such as the average diffusion coefficient, the average mobility, the mean free path, and the electron scattering mechanisms in very wide temperature range. The generalized expressions for basic electron transport characteristics, which are valid for materials both with non-degenerate and degenerate electron gas, are presented. It is well known that electrons obey the Pauli principle and they are described by the Fermi-Dirac statistics, and using the total density of free valence electrons for estimation of transport characteristics of electrons in materials with degenerate electron gas is unacceptable with respect to Fermi-Dirac statistics, because all these characteristics are determined by RM electrons near the Fermi level energy. An application of the classical statistics leads to colossal errors in the estimation of transport characteristics of electrons in materials with degenerate electron gas. It is shown that the Einstein's relation between the diffusion coefficient and drift mobility of RM electrons is held at any level of degeneracy of electron gas. The presented general expressions are applied for estimation of the carrier transport characteristics in the superconductor $YBa_2Cu_3O_{7-x}$ in the normal state, especially for description of the Hall-effect. It is shown that drift mobility of electrons in materials with degenerate electron gas can be tens or hundred times larger than the Hall mobility. The calculation results of the resistivity and other transport characteristics for elemental metals in temperature range from 1 K to 900 K are represented and compared with the experimental data.

Keywords: effective density of randomly moving electrons, electron diffusion coefficient, electron drift mobility, electron scattering, resistivity temperature dependence, mean free path, Hall effect of metals

INTRODUCTION

In this study it is attempted to call attention to the problems of charge carrier transport interpretation of metals, superconductors in the normal state, and semiconductors with highly degenerate electron gas. An application of the free electron transport in terms of the Fermi-Dirac statistics is presented, especially, the careful study is addressed to an application of the effective density of randomly moving (RM) electrons for description of the basic electron transport characteristics of materials with degenerate electron gas. Here, the term of the "effective density of randomly moving electrons" describes that part of free electrons near the Fermi level energy that can be scattered, and can change their energy due to the action of external fields. At a weak electric field the electrical conductivity of homogeneous materials is determined by free RM charge carriers and by their drift mobility. Though it is known that electrons obey the Pauli principle, and are described by the Fermi distribution function, but often, the classical (Boltzmann) statistics are applied to these materials. An aplication of the classical statistics leads to colossal errors in estimation of the effective density of RM electrons, and their kinetic coefficients for materials with degenerate electron gas.

Fermi-Dirac statistics of electrons allows us to explain the experimental results of the heat capacity of free electrons in metals: why metals and insulators have around the same heat capacity. The resolution of this paradox is one of the greatest successes of Sommerfeld's model (Sommerfeld and Bethe 1967). The main conclusion from this model was that only a small part of electrons move randomly, with their energy close to the Fermi level energy, and the other part of electrons where the energy is well below the Fermi level energy cannot change their energy because all neighbor energy levels are occupied. But the total density of the valence electrons n, is traditionally used for description of the electrical conductivity σ of metals and superconductors in the normal state by using expression $\sigma = q^2 n\tau/m^*$, where q, τ, and m^* are, respectively, the electron charge, the average electron relaxation time, and its effective mass. The main weakness of Sommerfeld's model is that it does not exactly define the effective density of the free RM electrons, which can be scattered, and which can be affected by external fields. Understanding that Sommerfeld's model is based on the spherical Fermi surface, there are also uncertainties in the determination of both the density-of-states (DOS) in conduction band at the Fermi surface, and the Fermi energy E_F, because the Fermi surfaces for most of the metals are not spherical (Cracknell and Wong 1973). Thus, an adequate definition of the effective density of the free RM electrons in materials with degenerate electron gas is needed. This study mainly addresses the stochastic description of the effective density of

RM electrons, to the estimation of their basic transport characteristics, and to the investigation of their scattering mechanism in materials with degenerate electron gas.

STOCHASTIC DESCRIPTION OF FREE ELECTRON PROPERTIES IN METALS

The Effective Density of Free Randomly Moving Electrons

It is well known that in materials, an electron with energy E is described by Fermi distribution function as

$$f(E) = 1/\{[1 + \exp[(E - \eta)/kT]\}, \tag{1}$$

where η is the chemical potential, k is the Boltzmann's constant, and T is the absolute temperature. This expression specifies the probability that an energy level E is occupied by an electron. The chemical potential η and the Fermi level energy E_F are related by the following relation (Ashcroft and Mermin 1976):

$$\eta = E_F[1 - (1/3) \cdot (\pi kT/2E_F)^2]; \tag{2}$$

the difference between the quantities η and E_F is only about 0.01 %, even at room temperature. Therefore, for the calculation of various quantities we shall use the Fermi distribution function in the following form:

$$f(E) = 1/\{[1 + \exp[(E - E_F)/kT]\}. \tag{3}$$

The total density n of the free electrons in the conduction band is then described by an integral over the density-of-states (DOS) in the conduction band as:

$$n = \int_0^\infty g(E)f(E)dE. \qquad (4)$$

The electrical conductivity depends not only on the DOS for electrons in the conduction band $g(E)$ and the Fermi distribution function $f(E)$, but it also depends on the probability $f_1(E) = 1-f(E)$ that any electron with the particular energy E at a given temperature T can be thermally excited (scattered) or can change its energy under the influence of external fields.

Thus, the effective density of RM electrons n_{eff} which take part in random motion and in conductivity is determined by probability $h(E) = f(E)[1-f(E)]$. This idea first has been pointed out in the work of (Palenskis, Juškevičius and Laucius 1985). The probability $h(E)$ dependence on energy of electrons near the Fermi level energy $E_F = 4$ eV (typical Fermi energy for good metal conductors) is shown in Figure 1 at temperature 295 K. There it can be noted that probability $h(E)$ also takes into account the Pauli's exclusion principle. The electrons which are described by probability $h(E)$ can move completely randomly, and can be scattered and change their energy under influence of external fields.

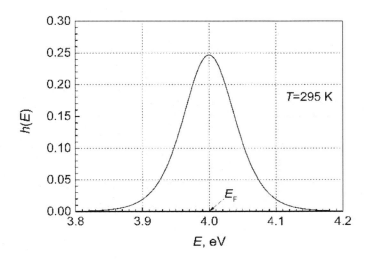

Figure 1. The probability $h(E) = f(E)[1-f(E)]$ distribution that electron with energy E is available to change its energy near the Fermi energy $E_F = 4$ eV at $T = 295$ K.

Therefore, the effective density of the RM electrons n_{eff} can be described as (Palenskis 2013; 2014a):

$$n_{eff} = \int_0^\infty g(E)f(E)[1 - f(E)]dE \tag{5}$$

or

$$n_{eff} = \int_0^\infty g(E)(-\partial f(E)/\partial E)dE. \tag{6}$$

From Eqs. (5) and (6) it follows that the term $(-\partial f(E)/\partial E)$ is the probability density function $p(E)$ of the RM electrons distribution on energy E:

$$p(E) = -\partial f(E)/\partial E = f(E)[1 - f(E)]/kT; \tag{7}$$

then the integral probability distribution function is:

$$F(E) = \int_0^E p(E_1)dE_1 = 1 - f(E). \tag{8}$$

The probabilities $p(E)$ and $F(E)$ meet the requirements of the probability theory. Its dependence on the energy of electrons in the surroundings of the Fermi energy $E_F = 4$ eV at T = 295 K are presented in Figure 2. The dash-line restricted rectangular area is equal to the area below the curve $p(E)$, and equal to 1. An effective width of the probability density function $p(E)$ distribution on energy of RM electrons is equal $\Delta E_{eff} = 4kT$. Thus, the effective density of the RM electrons is a random quantity with the probability density $p(E)$, while the total density of the free electrons in metals is constant and defined by the sum of valence electron density.

Illustration of Eqs. (3)–(5) is presented in Figure 3 for composite DOS: the light grey area under the curve $g(E)f(E)$ represents the total density of free electrons n in the conduction band, and the dark area under the curve $g(E)f(E)[1-f(E)]$ represents the effective density n_{eff} of the RM electrons.

From this figure it is seen that the effective density of the RM electrons is many times smaller than the total density of the free electrons.

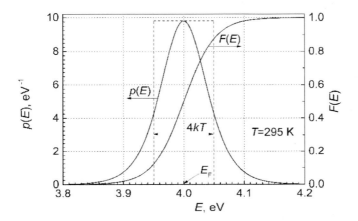

Figure 2. The probability density function $p(E)$ and the integral probability distribution function $F(E)$ dependences on energy of the RM electrons in the surroundings of the Fermi energy $E_F = 4$ eV at temperature $T = 295$ K.

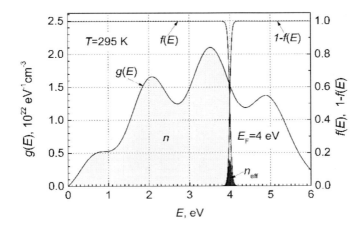

Figure 3. Illustration of the composite DOS $g(E)$ (left scale), the total density n of the valence electrons (light grey area, Eq. (4)), and the effective density n_{eff} of the RM electrons (black area, Eq. (5)) at the Fermi energy $E_F = 4$ eV and temperature $T = 295$ K. The Fermi functions $f(E)$ and $[1-f(E)]$ there also are presented (right scale).

Now let us evaluate the RM electron density [Eq. (5)] in homogeneous materials at any electron gas degree of degeneracy. For materials with non-degenerate electron gas, the probability $f_1(E)=[1-f(E)]\cong 1$, because $f(E)<<1$, and so, all electrons in the conduction band can be scattered, and can take part in conduction:

$$n_{\text{eff}}=n=\int_0^\infty g(E)f(E)dE. \tag{9}$$

This is the case when classical statistics are applicable. In the case of high degeneracy of the electron gas and considering that function $f(E)[1-f(E)]$ has a sharp maximum at $E=E_F$; then Eq (5) can be rewritten as:

$$n_{\text{eff}} = g(E_F)kT \ll n. \tag{10}$$

This relation shows that the effective density of the RM electrons in metals is completely determined by the DOS at Fermi energy $g(E_F)$, and it is also proportional to temperature. The DOS $g(E_F)$ can be obtained from experimental data of the electronic heat capacity (Ashcroft and Mermin 1976; Kittel 1976):

$$C_{\text{el}} = (\pi^2/3)g(E_F)k^2T = \gamma T. \tag{11}$$

where γ is coefficient of the electronic heat capacity. According to Sommerfeld's model based on the spherical Fermi surface, the DOS at Fermi surface can be expressed by the total density of the valence electrons n in the conduction band by the following relation (Ashcroft and Mermin 1976):

$$g(E_F) = (m^*/\hbar^2 \pi^{4/3})(3n)^{1/3}; \tag{12}$$

or by

$$g(E_F) = (3/2)n/E_F, \qquad (13)$$

where ℏ is the Plank's constant, m^* is the effective mass of the electron, E_F is the Fermi energy.

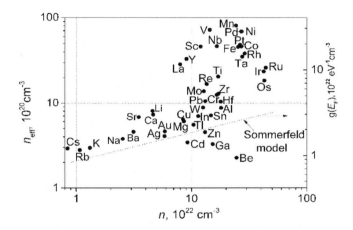

Figure 4. The DOS at Fermi energy $g(E_F)$ (right scale), and the effective density of RM electrons n_{eff} [Eq. (10), left scale at T = 295 K] distribution on the total density of the valence electrons n in the conduction band for elemental metals. The dash line is under Sommerfeld's model [Eq. (12)]. Data $g(E_F)$ are calculated by Eq. (11).

As it is seen from Figure 4, the Sommerfeld model for many of metals gives large errors. Moreover, in the cases when $g(E_F)$ values are close to that obtained by Eq. (12), as in the case for the Tl and Zn, the Fermi surfaces of these metals are completely non-spherical (Cracknell and Wong 1973). The presented results show that neither Eq. (12) is useful for definition of DOS of electrons at the Fermi surface of metals, nor is Eq.(13) useful for the determination of the Fermi energy. The application of Eqs. (12) and (13) lead to the Drude formula for conductivity in the form $\sigma = q^2 n \tau / m^*$, but it is incorrect for metals, because not all valence electrons, as it has been proven earlier, can randomly move and be affected by external field. As it has been pointed out, the correct values of the DOS at the Fermi surface for metal are obtained from the experimental measurement results of electronic heat capacity. An advantage of these data $g(E_F)$ is that for normal simple metals, they do not depend on temperature. The electronic heat values of all simple metals are measured and tabulated (Kitel 1976; Lide 2003–2004).

In addition, the application of Sommerfeld Eqs. (12) and (13) to the description of the electronic thermal conductivity λ as (Alloul 2011; Kittel 1976):

$$\lambda = (1/3) l_F v_F C_{el} = (\pi^2/9) k^2 g(E_F) T v_F^2 \tau_F \tag{14}$$

also gives the wrong result for electronic thermal conductivity:

$$\lambda = (\pi^2/6) n k^2 v_F^2 \tau_F T/E_F = (\pi^2/3) n k^2 T \tau_F / m^*, \tag{15}$$

though this model gives the correct Wiedeman-Franz law:

$$\lambda/(\sigma T) = (\pi^2/3)(k^2/q^2), \tag{16}$$

because both electronic electrical and thermal conductivities have the same incorrect parameter n, and the ratio $\lambda/(\sigma T)$ does not depend on n.

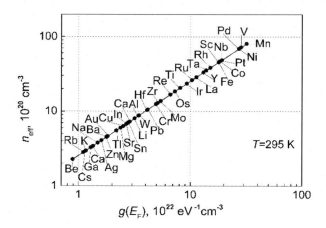

Figure 5. The relation between the effective density n_{eff} of the RM electrons at $T = 295$ K and the DOS of electrons on the Fermi surface $g(E_F)$ (the same as in Figure 4).

Study of the Transport of Charge Carriers in Materials ... 133

The distribution of the effective density of RM electrons versus the DOS of electrons at Fermi surface for elemental metals is presented in Figure 5. The volume estimated for one RM electron with the Fermi velocity is equal to:

$$V_{1el} = 1/n_{eff} = 1/[g(E_F)kT]. \tag{17}$$

The largest volume for a RM electron belongs to beryllium, while for manganese or vanadium is about 30 times smaller. This volume increases with temperature decreasing. It is interesting to point that more conductive (alkali and noble) metals have small effective densities of RM electrons.

Thus, the probability density function completely defines the effective density of the RM electrons in metals and other homogeneous materials.

The General Description of the Electrical Conductivity of Metals

Considering that the effective density of RM electrons n_{eff} in metals is not equal to the total density of the free valence electrons in the conduction band, the classical conductivity description by expression $\sigma = q^2 n\tau/m^*$ is completely unacceptable. Usually a general description of the electrical conductivity of conductive materials is obtained by solving the Boltzmann's kinetic equation (Abrikosov 1987; Dugdale 2010; Seeger 1984), and is presented as:

$$\sigma = 2q^2/(3m^*kT) \int_0^\infty \tau(E)g(E)f(E)[1 - f(E)]EdE, \tag{18}$$

where $\tau(E)$ is the relaxation time of electrons with kinetic energy E in conduction band. This expression after simple transfer can be written in the following form (Dugdale 2010):

$$\sigma = (q^2D/kT) \int_0^\infty g(E)f(E)[1 - f(E)]dE, \tag{19}$$

or as (Palenskis 2013):

$$\sigma = q^2 D n_{\text{eff}}/kT, \tag{20}$$

where $D=(1/3)<v^2> \cdot <\tau>$ is the diffusion coefficient of the RM electrons, v is their kinetic velocity, and n_{eff} is described by Eq. (5). The electrical conductivity also can be described as (Bisquert 2008; Dugdale 2010):

$$\sigma = q^2 D (\partial n/\partial \eta)_T, \tag{21}$$

where η is the chemical potential. There, all presented descriptions of conductivity by Eqs. (18)–(21) are equivalent, because they can all be derived, one from the other. Moreover, they are valid in all cases for homogeneous materials with one type of charge carriers (electrons or holes) at any electron gas degree of the degeneracy, and it confirms that the electrical conductivity for all metals is determined by the effective density of the RM charge carriers, but not by the total density of the free valence electrons in the conduction band. On the other hand, the electrical conductivity can be described as:

$$\sigma = q n_{\text{eff}} \mu_{\text{drift}}, \tag{22}$$

where μ_{drift} is the drift mobility of the RM charge carriers. Eqs. (20) and (22) follow the fundamental Einstein's relation:

$$D/\mu_{\text{drift}} = kT/q, \tag{23}$$

which is true for all homogeneous materials with one type of RM charge carriers at any density. Therefore, it is proved that Einstein's relation is valid not only for semiconductors with non-degenerate electron gas, but also for metals, superconductors in the normal state, and for very highly doped semiconductors. The right side of Eq. (23) represents the thermal potential $\varphi_T = kT/q$. In principle, Einstein's relation shows that the current components caused by drift and diffusion at equilibrium compensate one another, and the resultant average total current is always equal zero: it does not depend on the degeneration of electron gas.

Now let us treat Eq. (20) for both cases of non-degenerate and highly degenerate electron gas. In the case of non-degenerate electron gas, $n_{\text{eff}} = n$, and the electrical conductivity can be described as:

$$\sigma = q^2 Dn/kT = qn\mu_{\text{drift}}. \tag{24}$$

This is the case when classical statistics are applicable. From Eqs. (20) and (22), follow such general drift mobility description (Palenskis 2013; 2014a):

$$\mu_{\text{drift}} = qD/kT = (q < \tau >/m^*) \cdot < E >/(3kT/2), \tag{25}$$

where $< E >= m^* < v^2 >/2$ is the average kinetic energy of the RM charge carriers. It is a fundamental relation for drift mobility of RM charge carriers in homogeneous materials with one type of charge carriers (electrons or holes). The relation (25) is valid at any degeneracy degree of electron gas. The factor:

$$\alpha_\varepsilon =< E >/(3kT/2) \tag{26}$$

shows how many times the average kinetic energy $< E >$ of the RM charge carriers is larger than the thermal energy $3kT/2$. For non-degenerate electron gas $\alpha_\varepsilon = 1$ (the case of classical statistics), and $\mu_{\text{drift}} = q < \tau >/m^*$, but for metals, superconductors in the normal state $\alpha_\varepsilon = 2E_F/(3kT) \gg 1$, i. e. the drift mobility of the RM electrons in metals at room temperature is about one hundred times larger than it follows from the classical expression $\mu = q < \tau >/m^*$.

The Eq. (20) for metals can be rewritten as:

$$\sigma = 1/\rho = q^2 g(E_F)D = (1/3)q^2 g(E_F)v_F^2 \tau_F, \tag{27}$$

where ρ is the resistivity of metal; v_F and τ_F are, respectively, the electron velocity and its relaxation time at the Fermi surface. This expression is well known for metals (Abrikosov 1987; Kittel 1976; Ziman 1972). In Figure 6, there is shown the relation of the conductivity of elemental metals with DOS at the Fermi surface at $T = 295$ K. It shows a tendency of conductivity decreasing with DOS increasing as $g^{1/2}(E_F)$ (dash line). The proportionality of the conductivity to $g^3(E_F)$ (a dash line in Figure 6) follows from Sommerfeld's model according to Eq. (12). So, Sommerfeld's model, based on the spherical Fermi surface of metals, in principle, cannot explain the data shown in Figure 6. From Eq. (27), it follows that the scattering of the conductivity results on DOS is due to different diffusion coefficients of the RM electrons in various metals.

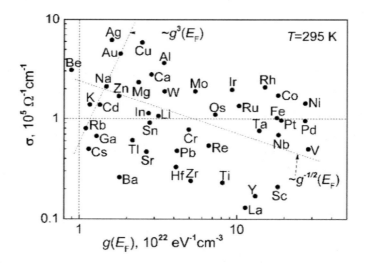

Figure 6. Distribution of the conductivity for elemental metal versus DOS at the Fermi surface at $T = 295$ K. The conductivity data are taken from the table in the book (Kittel 1976); $g(E_F)$ are the same as in Figure 4.

Considering the effective density of the RM electrons, the electronic thermal conductivity can be described as:

$$\lambda = (1/3)l_F v_F C_{el} = (\pi^2/3)Dg(E_F)k^2T = (\pi^2/3)n_{eff}kD, \qquad (28)$$

and from Eqs. (27) and (28) follows the same ratio (Wiedeman-Franz law) for the electronic thermal and electrical conductivities [Eq. (16)], but in this case, both electronic thermal and electrical conductivities are well determined.

Diffusion Coefficient and Drift Mobility of Electrons in Metals

Eq. (27) lets us simply estimate the diffusion coefficients of the RM electrons for various metals, because the electronic heat capacities for all elemental metals are known. The diffusion coefficient of the free RM electrons is described as:

$$D = \sigma/[q^2 g(E_F)]. \tag{29}$$

and then the drift mobility from Eq. (24) as:

$$\mu_{drift} = \sigma/(qn_{eff}) = \sigma/[qg(E_F)kT]. \tag{30}$$

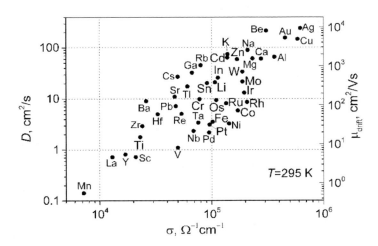

Figure 7. Relationship between conductivity, diffusion coefficient [Eq. (29)] and drift mobility [Eq. (30)] of the RM electrons for elemental metals at $T = 295$ K. The conductivity data are taken from the table in the book (Kittel 1976).

The average values of the drift mobilities and diffusion coefficients of the RM electrons for elemental metals at room temperature (295 K) determined, respectively, by Eqs. (29) and (30) are presented in Figure 7.

Considering that at a given temperature, the ratio $D/\mu_{drift} = kT/q$ is constant, the spread of the data in this figure is caused by different values of DOS of electrons at Fermi energy for different metals.

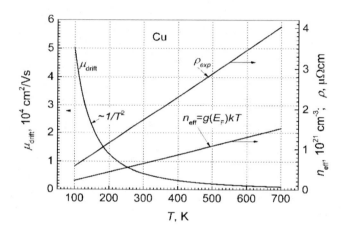

Figure 8. The resistivity, drift mobility [Eq. (30)] and effective density of RM electrons [Eq. (10)] dependences on temperature for copper over the Debye's temperature.

As seen from the Figure 7, the drift mobility of the RM electrons for some of metals is very high. For example, the drift mobility of RM electrons in silver at room temperature is about 10^4 cm²/(Vs), but the Hall mobility for it is only about 68 cm²/(Vs) (Blakemore 1985). Illustration of the changes of resistivity components in temperature range, where resistivity changes linearly with temperature, for copper is presented in Figure 8. It is important to note that with temperature increasing, it increases both the resistivity and the effective density of the RM electrons, because $n_{eff} = g(E_F)kT$, while in the Drude and Sommerfeld models the total mobile electron density n is constant. The other important difference: in these models it is confirmed that drift mobility changes with temperature as $1/T$, but really it changes as $1/T^2$ (Eq. (25)). The presented results show

Study of the Transport of Charge Carriers in Materials ... 139

that Sommerfeld's model, based on the spherical Fermi surface, can explain why the electronic heat capacity is small, but it cannot define the effective density of RM electrons and their drift mobility, and that Einstein's relation between the diffusion coefficient and drift mobility of RM electrons are valid for any degeneration of electron gas.

Thereinafter, we will present what information one can get from Hall-effect measurement based on the Fermi distribution of electrons.

Investigation of the Hall Effect of Metals

Usually, in measurement of the Hall effect, a rectangular plate with dimensions $d \times w \times l$ (thickness×width×length) is used. The magnetic field is directed along z-axis, and the direct current flows only along x-axis, and current density $j_y = j_z = 0$. The originated electric field in y-direction $E_y = E_H = U_H / w$ (here U_H is the Hall voltage) compensates the deflection change carrier due to the Lorentz force. The Hall coefficient R_H is defined as (Smith 1978; Sólyom 2009):

$$R_H = E_y / (j_x B_z) = E_y / (\sigma E_x B_z), \tag{31}$$

where j_x is the current density along x-axis, B_z is the magnetic flux density in z-direction, σ is the conductivity of the sample material with one type of charge carriers. Instead of the Hall coefficient, the Hall effect is sometimes characterized by the Hall angle φ. For small, both direct current density and magnetic flux density the $\tan\varphi \cong \varphi$ the Hall angle is described as:

$$\varphi = E_y / E_x = E_H / E_x. \tag{32}$$

On the other hand, the Hall angle can be expressed by cyclotron angular frequency $\omega_c = q B_z / m^*$ of the charge carrier moving perpendicular to the direction of the uniform magnetic field (Smith 1978):

$$\varphi = \omega_c <\tau> (<\tau^2>/<\tau>^2) = (qB_z/m^*) <\tau> r_H, \tag{33}$$

where $<\tau>$ is the electron mean free flight time, $r_H=<\tau^2>/<\tau>^2$ is the Hall factor which depends on the scattering mechanism of the charge carriers, and changes from 1 to 2; here quantity q also includes the charge polarity sign. It is notable that ω_c is independent of the radius and velocity, and therefore, independent of the particle kinetic energy: all particles with the same charge-to-mass ratio rotate around magnetic field lines with the same cyclotron angular frequency. So, electrons with the same effective mass moving with thermal velocity in material with non-degenerate electron gas, and electrons moving with the Fermi velocity in metals have the same ω_c. From Eqs. (31)–(33), it follows that:

$$E_y/(E_x B_z) = [q <\tau>/m^*]r_H = \mu_H = R_H \sigma, \tag{34}$$

considering that quantity μ_H has the similar expression as for mobility of the charge carriers in semiconductors with non-degenerate electron gas, it is named as Hall mobility. This expression is valid at any degeneration degree of the electron gas with one type of RM charge carriers. The quantity R_H has a negative sign for electrons and positive sign for holes. Note: if one compensates E_y by external bias source to zero (as it is usually done in the measurements of the Hall voltage), but it has no influence on both the cyclotron frequency and drift mobility of charge carriers of the material.

In general case, the conductivity of the homogeneous material with one type of the charge carriers from Eqs. (22), (25) and (26) can be described as:

$$\sigma = qn_{eff}\mu_{drift} = \alpha_\varepsilon q^2 n_{eff} <\tau>/m^* \tag{35}$$

where $\alpha_\varepsilon =<E>/(3kT/2)$. Then from Eqs. (34) and (35) follows such general expression for the Hall coefficient:

$$R_H = r_H/(\alpha_\varepsilon q n_{eff}). \tag{36}$$

For homogeneous material with non-degenerate electron gas $\alpha_\varepsilon = 1$ and the drift mobility can be expressed as:

$$\mu_{drift} = \mu_H r_H = r_H q <\tau>/m^*. \tag{37}$$

Then the Hall coefficient has the usual form:

$$R_H = r_H/(q n_{eff}) = r_H/(qn). \tag{38}$$

Thus, for materials with one type of the non-degenerate electron gas can, from the Hall effect, measure data to determine the free charge carrier density n and their drift mobility μ_{drift}.

In the case of metals and other materials with the highly degenerate electron gas $r_H = 1$, the electron relaxation time is equal to the electron relaxation time on the Fermi surface: $<\tau> = \tau_F$. Then the drift mobility of the RM charge carriers is described as:

$$\mu_{drift} = (q\tau_F/m^*)[E_F/(3kT/2)] = Dq/kT, \tag{39}$$

where $D = v_F^2 \tau_F/3$ is the diffusion coefficient of the electrons. It is seen that the drift mobility of RM electrons is many times larger than the Hall mobility, and does not depend on the effective mass of the charge carriers. Then from Eq. (36), it follows that the Hall coefficient for metals and materials with the degenerate electrons is described (Palenskis 2015) as:

$$R_H = (3/2)(1/[qE_F g(E_F)]. \tag{40}$$

This relation is valid only for metals with single type of charge carriers. For the ideal spherical Fermi surface, the product $E_F g(E_F) = 3n/2$ [Sommerfeld model, Eq. (13)], and one can get the known relation:

$$R_H = 1/(qn). \tag{41}$$

Considering that Fermi surfaces for most of metals are completely not spherical (Cracknell and Wong 1973), the estimation of the total density of the free valence electrons for metals by the Eq. (41) gives very large errors. It also, for real metals, leads to inaccurate estimation of the Fermi energy from both the DOS at the Fermi surface and the total density of the free electrons in the conduction band (Figure 3). Besides, the total density of the free valence electrons can simply be estimated from the metal atom valency and the density of atoms in metal. Considering that for most metals, the Fermi surfaces are not spherical, the estimations by using spherical Fermi surfaces (Sommerfeld's model) give inaccurate results.

The Hall coefficient measurement results show that it can take not only negative but also a positive sign: it means that in metals, as in semiconductors, the electrons and holes take part in the conduction. The sign depends on the quasiparticle effective mass. The effective mass m^* of the free electrons of a metal in one-dimentional case of $\mathbf{k}$-space is given by (Linde 1964)

$$m^* = \hbar^2/(\partial^2 k/\partial k^2), \tag{42}$$

where $\mathbf{k}$ is the wave-vector of the electrons at the Fermi surface. The effective mass reflects a curvature of the energy surfaces in different sites of the Fermi surfaces. The investigation of the Fermi surfaces of metals (Cracknell and Wong 1973) shows that in particular areas of the Fermi surface there are electron-like and hole-like areas.

Therefore, in research of the Hall effect and conductivity of metals, one has to account that effective densities of both electrons n and holes p, and their drift mobilities μ_n and μ_p (Markiewicz 1988; Palenskis 2015; Ziman 1972). Then the conductivity can be described as:

$$\sigma = \sigma_n + \sigma_p = qn\mu_n + qp\mu_p = q^2 g_{\text{total}}(E_F)D \tag{43}$$

Study of the Transport of Charge Carriers in Materials ...   143

where D is the diffusion coefficient of the free RM charge carriers. Figuring that at equilibrium conditions, the Fermi energy is the same in all sites of the Fermi surface, it has been considered that all RM charge carriers are moving randomly with the same average Fermi velocity and they have the same diffusion coefficient D and the same average relaxation time. In the case of different diffusion coefficients of electrons and holes, in some sites the excess charge could accumulate, which cannot take place. Then we can write the following relations:

$$n = n_{\mathrm{eff}} = g_n(E_\mathrm{F})kT \tag{44}$$

is the effective density of RM electrons at the Fermi surface at temperature T, $g_n(E_\mathrm{F})$ is the electron-like effective DOS at the Fermi surface;

$$p = p_{\mathrm{eff}} = g_p(E_\mathrm{F})kT \tag{45}$$

is the effective density of RM holes at the Fermi surface at temperature T, $g_p(E_\mathrm{F})$ is the hole-like effective DOS at the Fermi surface;

$$|\mu_n| = |\mu_p| = |\mu_{\mathrm{drift}}| = |q|v_\mathrm{F}^2 \tau_\mathrm{F}/(3kT) \tag{46}$$

is the drift mobility, respectively, for the electrons there is index n, and for holes – index p. From Eqs. (43)–(46) the conductivity of metals with two types of the free RM charge carriers can be expressed as:

$$\sigma = (1/3)q^2 v_\mathrm{F}^2 \tau_\mathrm{F} g_{\mathrm{total}}(E_\mathrm{F}), \tag{47}$$

where $g_{\mathrm{total}}(E_\mathrm{F}) = g_n(E_\mathrm{F}) + g_p(E_\mathrm{F})$ is the total DOS at the Fermi surface. The Hall coefficient for two types of the free RM charge carriers (electrons and holes) is described (Dresselhaus 2001a; Markiewicz 1988; Ziman 1972) as:

$$R_{\text{H2}} = (R_{\text{H}p}\sigma_p^2 + R_{\text{H}n}\sigma_n^2)/\sigma^2 = (qp_{\text{eff}}\mu_{\text{H}p}\mu_{p\text{drift}} - qn_{\text{eff}}\mu_{\text{H}n}\mu_{n\text{drift}})/\sigma^2 \tag{48}$$

where $R_{\text{H}i}\sigma_i = \mu_{\text{H}i}$, which is valid for a separate free charge type. As shown in the work (Palenskis 2015), there also the following valid relations:

$$R_{\text{H}n} = -1/(q\alpha_\varepsilon n_{\text{eff}}) = -3/[2qg_n(E_\text{F})E_\text{F}] \tag{49}$$

is the Hall coefficient caused by free RM electrons;

$$R_{\text{H}p} = 1/(q\alpha_\varepsilon p_{\text{eff}}) = 3/[2qg_p(E_\text{F})E_\text{F}] \tag{50}$$

is the Hall coefficient caused by free RM holes;

$$\mu_{\text{H}n} = R_{\text{H}n}\sigma_n = -q\tau_\text{F}/m \tag{51}$$

is the Hall mobility for free RM electrons, and

$$\mu_{\text{H}p} = R_{\text{H}p}\sigma_p = q\tau_\text{F}/m \tag{52}$$

is the Hall mobility for free RM holes. Considering that $R_{\text{H}n}$ has a negative sign, and $R_{\text{H}p}$ has a positive sign, they partly compensate one another. Then Eq. (48) can be rewritten as:

$$R_{\text{H2}} = (qp\mu_{\text{H}p}\mu_p - qn\mu_{\text{H}n}\mu_n)/\sigma^2 \tag{53}$$

and after inserting in Eq. (53)), the conductivity [Eq. (47)], the effective densities of RM charge carriers [Eqs. (44) and (45)], their Hall [Eqs. (51) and (52)] and drift mobilities [Eq. (46)], we obtain the following expression:

$$R_{\text{H2}} = |R_{\text{H0}}|[1 - 2g_n(E_\text{F})/g_{\text{total}}(E_\text{F})], \tag{54}$$

where

$$|R_{H0}| = |3/[2qE_Fg_{total}(E_F)]| = |3/[qmv_F^2 g_{total}(E_F)]| \qquad (55)$$

is the expression of the Hall coefficient for a sample only with one type of the free RM charge carriers. From Eq. (54) it is seen, if electron-like DOS at the Fermi surface is larger than one of hole-like DOS, then the Hall coefficient has negative sign; in the opposite case it has positive sign. When the electron-like DOS is approximately equal to hole-like DOS, the Hall coefficient is near zero. Therefore, for interpretation of the Hall measurement data, one must include both electrons and holes. More features about the Hall effect will be presented in the section on high-temperature superconductor properties in the normal state.

Scattering and the Electron Mean Free Path in Metals

The resistivity of metals above the Debye's temperature increases linearly with temperature, and this behavior is usually explained by scattering due to lattice vibrations, *i. e.* due to an increase of the intensity of the thermal vibration of atoms of the lattice, and that it causes the increase of the electron-phonon scattering cross-section (Abrikosov 2017; Ashcroft and Mermin 1976; Kaveh and Wiser 1984; Lundstrom 2014; Rositer 2014; Schulze 1967; Sondheimer 2001; Wilson 1958; Ziman 1972; 2001). This model cannot explain why the electron real mean free path is about one or two orders larger than the interatomic distance. There, in all cases it is proposed that all valence electrons in the conduction band can randomly move, and can be scattered; but this preposition, as it was earlier shown, is wrong and contradicts the Fermi-Dirac statistics for electrons. Considering that the free RM electron density increases as the temperature increases ($n_{eff} = g(E_F)kT$), another explanation model is needed.

A very important parameter characterizing the scattering mechanism of RM charge carriers is their mean free path. According to quantum mechanics, a periodic potential energy originates from the periodic lattice of the ions and the average free electron interaction. The free electrons as Bloch waves can freely move in the ideal periodic lattice of the metal with periodic distribution of the potential energy without any scattering by lattice ions (Ashcroft and Mermin 1976; Ziman 2001). It means that the ions vibrations only slightly influence the periodicity of potential distribution in the periodic lattice. The ideal periodicity of the potential energy of the perfect lattice also causes the periodical distribution of the charge density. The scattering of the free RM electrons can only be in the spots where there are distortions of the periodicity of the potential energy of the ideal lattice structure. The distortions of the periodic potential distribution in real crystals are caused by the presence of impurities, interstitials, vacancies, dislocations, limits of the grains, and by the surface of tested samples that create a resistance to current flow. The electron mean free path l_F due to the named defects almost does not depend on temperature, and their dominance appears at very low temperatures, while in very wide temperature range the mean free path depends on temperature: at temperature above the Debye's temperature $l_F \sim T^{-1}$, and at temperature below the Debye's temperature it usually changes as T^{-5} (Sondheimer 2010).

The average charge density in the volume $V_{at} = 1/N_{at}$ equals zero due to electrical neutrality of the material (here N_{at} is the density of material atoms). The free electrons in ideal perfect lattice move in such a way that some of them cause the local fluctuation of the charge density due to phonon stimulated escape of the electron from particular ion to long distances of the order of the free path, and it produces the local distortion of the potential periodicity due to not completely screening this ion by electrons of adjacent ions. Such local distortion in the periodicity of the potential energy (or of the charge density) is named by electronic defects (Wert and Thomson 1964).

The scattering of free RM electrons is inelastic due to the fluctuations of the energy of the free RM electrons. The average kinetic energy $<E>$ of the free RM electron in metals is:

$$< E >/n_{\text{eff}} = [\textstyle\int_0^\infty Eg(E)p(E)dE]/[\textstyle\int_0^\infty g(E)p(E)dE] = E_{\text{F}}, \qquad (56)$$

i.e., the average energy of the RM electron at the Fermi surface is equal to the Fermi level energy. Now we estimate the free RM electron energy fluctuation variance:

$$\langle(E-< E >)^2\rangle = \langle(E - E_{\text{F}})^2\rangle ==\textstyle\int_0^\infty (E - E_{\text{F}})^2 g(E)p(E)dE$$
$$= g(E_{\text{F}})(kT)^3 \textstyle\int_0^\infty (\varepsilon - \varepsilon_{\text{F}})^2 p(\varepsilon)d\varepsilon \cong(\pi^2/3)g(E_{\text{F}})(kT)^3. \qquad (57)$$

where $\varepsilon = E/(kT)$. From comparison of this variance expression with Eq. (11) for electronic heat capacity we can write that:

$$\langle(E-< E >)^2\rangle = kT^2 C_{\text{el}}. \qquad (58)$$

This relation shows that the variance of the free RM electrons is described in the same way as for RM molecules in gas (Kittel 1969).

There we want to point out that lattice ion vibrations play another role, than it has been prescribed by (Abrikosov 2017; Ashcroft and Mermin 1976; Kaveh and Wiser 1984; Lundstrom 2014; Rositer 2014; Schulze 1967; Sondheimer 2001; Ziman 1972, 2001). With a temperature increase, the thermal vibration of lattice ions stimulated the increase of the density of the free RM electrons ($n_{\text{eff}} = g(E_{\text{F}})kT$), which produce the same density of the local distortion spots (electronic defects) $N_{\text{eff}} = n_{\text{eff}} = g(E_{\text{F}})kT$ from the ideal distribution of the periodicity of the potential energy (or the charge density) due to the escape of free RM electrons from particular ions to the long distance of order of the electron mean free path (Palenskis and Žitkevičius 2018; 2020). These electronic defects cause the scattering of the free electrons at the Fermi surface.

The intensity of scattering of the free RM electrons by electronic defects depends not only on the scattering cross-section of these defects, but it also depends on the ratio of the exchange of thermal energies between phonon and electron. The average thermal energy (Ashcroft and

Mermin 1976) of the free electrons estimated for one free RM electron is (Palenskis and Žitkevičius 2018):

$$E_{el1} = (\pi^2/6)g(E_F)\cdot(kT)^2/n_{eff} \cong 1.64kT. \tag{59}$$

The average phonon thermal energy at temperatures over Debye's temperature Θ is about $3kT$ (Kittel 1976; Ziman 1972) because all lattice waves are excited, but at a temperature range below the temperature Θ, only the phonons with low frequencies are exited. Accounting for the free RM electron energy fluctuations, which cause the excitation and annihilation of phonons, the average phonon thermal energy estimated for one phonon can be described as (Ziman 1972):

$$E_{ph1} = 3kT(T/\Theta)^4 \int_0^{\Theta/T} 4x^5/[(e^x - 1)(1 - e^{-x})]\,dx. \tag{60}$$

Then the ratio of the average thermal energies between the phonon to the electron can be expressed as:

$$E_{ph1}/E_{el1} \cong 1.83\eta(T/\Theta), \tag{61}$$

where

$$\eta(T/\Theta) = (T/\Theta)^4 \int_0^{\Theta/T} 4x^5/[(e^x - 1)(1 - e^{-x})]\,dx; \tag{62}$$

therefore, the phonon mediated resultant free RM electron scattering cross-section σ_{res} for metals can be described (Palenskis and Žitkevičius 2018) as:

$$\sigma_{res} = (E_{ph1}/E_{el1})\sigma_{eff} = \sigma_{res0}\eta(T/\Theta), \tag{63}$$

where $\sigma_{res0} = 1.83\sigma_{eff}$, and σ_{eff} is the effective electronic defect scattering cross-section without accounting the phonon mediation factor, and is caused by particular properties of the Fermi surface for every separate metal. The resultant scattering cross-section σ_{res} for metals can be found from the electron mean free path l_F dependence on the electronic defect density $N_{eff}=g(E_F)kT$:

$$l_F = v_F \tau_F = 1/(\sigma_{res}N_{eff}) = 1/[\sigma_{res}g(E_F)kT], \tag{64}$$

Then the average relaxation time τ_F of the free RM electrons can be described as:

$$\tau_F = 1/(\sigma_{res}N_{eff}v_F) = 1/[\sigma_{res}g(E_F)v_FkT] . \tag{65}$$

Considering that the Fermi velocity for elemental metals does not depend on temperature, from Eq. (65) it also follows that the resultant free electron scattering cross-section at $T>\Theta$ does not depend on temperature, because at these temperatures $\eta(T/\Theta) \cong 1$ is almost constant. Thus, the obtained Eq. (64) explains the large electron mean path comparing to the distance between atoms.

Electron Transport in Metals over the Debye's Temperature

From Eq. (64) it follows that the electron average free path dependence on temperature over the Debye's temperature is completely defined by the effective density of the electronic defects N_{eff}. Considering that N_{eff} increases with temperature increasing, it causes the electron mean free path to decrease as $1/T$.

Considering that the diffusion coefficient D of the free RM electrons in metals can be simply found from Eq. (27), by using the calculation data (Gall 2016) for electron mean path of large group of metals, and the electron mean path estimation results by dimensional effect (Chopra 1979), it is shown (Palenskis and Žitkevičius 2018) that in the linear resistivity, dependence on temperature range the electron mean free path for various metals is proportional to diffusion coefficient $D^{2/3}$. Then electron mean free path at $T>\Theta$ can be approximated as:

$$l_F(T) = 1.39[\gamma^{2/3}(T_0)](T_0/T), \tag{66}$$

where l_F is in nm, $\gamma(T_0) = D(T_0)/D_{\text{ref}}$ is the relative diffusion coefficient, where $D(T_0)$ is the free electron diffusion coefficient $D(T_0)$ in cm²/s at $T_0 = 295$ K, and $D_{\text{ref}} = 1 \text{ cm}^2/\text{s}$ is taken as the reference diffusion coefficient. The electron mean free path dependence on the relative diffusion coefficient is presented in Figure 9.

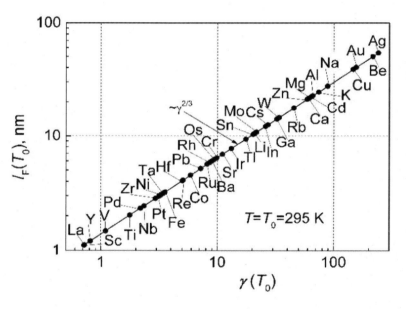

Figure 9. The relationship between electron mean free path and the relative diffusion coefficient of the free RM electrons for elemental metals at $T = 295$ K.

From Eqs. (63) and (64) it follows that resultant free electron scattering cross-section does not depend on temperature at $T>\Theta : \sigma_{\text{res}} = \sigma_{\text{res0}}$. It means that scattering cross-section at this temperature range is proportional to temperature (Ashcroft and Mermin 1976; Kaveh and Wiser

1984; Lundstrom 2014; Rositer 2014; Schulze 1967; Sondheimer 2001; Ziman 1972) is not right for RM electrons in metals. The resultant free electron scattering cross-section dependence on the DOS is presented in Figure 10.

The free RM electrons velocity at the Fermi surface can be simply evaluated by Eqs. (27) and (66) as:

$$v_F = 3D/l_F = 2.16\gamma^{1/3}(T_0), \tag{67}$$

where v_F is in units 10^7 cm/s, and it does not depend on temperature. This dependence on the relative diffusion coefficient is shown in Figure 11.

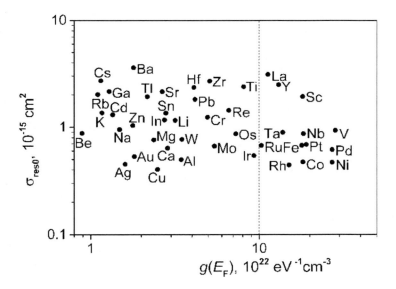

Figure 10. The resultant scattering cross-section σ_{res0} [Eq. (63)] distribution on DOS for elemental metals; DOS are the same as in Figure 6.

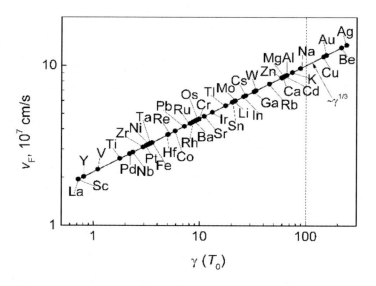

Figure 11. The Fermi velocity (calculated by $3D/l_F$) dependence on the relative diffusion coefficient of the free RM electrons for elemental metals.

It is interesting to see, how the Fermi velocity depends on the metal elements location in the Mendeleef's periodic table, i.e., on the number N_{tab} in this table (Figure 12). It is seen that large Fermi velocity has the Be and the Ib column metals (Cu, Ag, and Au) with cubic lattice, and the lowest one has the IIIb and IVb columns metals (Sc, V, Y, and La) with hexagonal lattice. The other regularities are not so well expressed.

Accounting for the Fermi velocity, we can estimate the Fermi energy of the RM electrons at the Fermi surface as $E_F = m_0 v_F^2/2$, where m_0 is the free electron mass. The obtained data are presented in Figure 13. It is seen that Fermi energy has a trend to decrease with DOS increasing. According to the Sommerfeld's model $g(E_F) \sim \sqrt{E_F}$ (Ashcroft and Mermin 1976). Therefore, the Sommerfeld's model is inapplicable for estimation of the Fermi energy of the free electrons, and it gives very large errors.

Study of the Transport of Charge Carriers in Materials ... 153

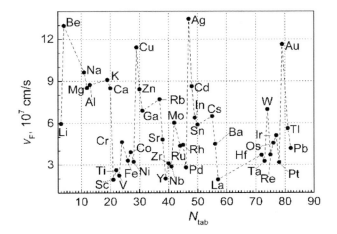

Figure 12. The Fermi velocity of the free RM electrons dependence on the metal element location number N_{tab} in the Mendeleef's periodic table (the dash lines are only used for connection between dots).

Considering that the relaxation time of the RM electrons τ_F on the Fermi surface is described as $\tau_F = l_F/v_F$, it also at $T>\Theta$ can be expressed by relative diffusion coefficient $\gamma(T_0)$ of the free RM electrons as:

$$\tau_F = 0.64[\gamma^{1/3}(T_0)](T_0/T), \qquad (68)$$

where τ_F is in units 10^{-14} s. The relaxation time of the free RM electron depends on the relative diffusion coefficient and is shown in Figure 14. The relaxation times at room temperature ($T = 295$ K) are distributed in the range from 6 fs to 40 fs.

According to (Devillers 1984), the average relaxation time of electrons for metals in the range of the linear resistivity dependence on temperature can be described as $\tau_F = \hbar/kT$, but such estimation is approximately applicable only for Al, Ca, K, Mg, Na and Zn, for which the relaxation time at room temperature is about 26 fs (Figure 14).

Therefore, we are presenting the simple expressions (Eqs. (66)–(68)) for estimation of the basic electron transport characteristics in metals at temperatures over the Debye's temperature only from resistivity (or conductivity) measurement results, because for all elemental metals the DOS at Fermi surface are known.

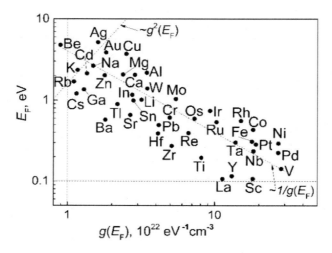

Figure 13. Fermi energy E_F dependence on DOS at the Fermi surface for elemental metals: the dash line $E_F \sim g^2(E_F)$ is according to Sommerfeld's model; the dash line $\sim 1/g(E_F)$ is a trend of the Fermi energy change with DOS increasing at the Fermi surface).

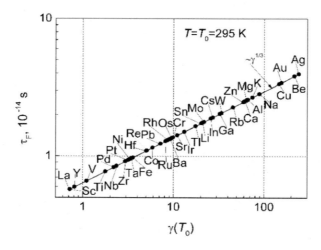

Figure 14. The free electron relaxation time (calculated as $\tau_F = l_F/v_F$) dependence on the relative diffusion coefficient of the free RM electrons for elemental metals at $T = 295$ K.

Phonon Mediated the Electron Transport below the Debye's Temperature

In Figure 15 there are plotted the measurement results of the electrical resistivity dependences on temperature in the temperature range from 1 K to 900 K for Au, Al, Ag, Sr, and W (Knowledge door 2019). The independent resistivity part on temperature below 10 K is the residual resistivity ρ_0 due to scattering caused by distortion of the potential energy periodicity by chemical and structural imperfections (defects) in the lattice of the presented samples. In this case the phonon mediated electron scattering by electronic defects scattering cross-section becomes negligible and scattering of electrons by impurities and defects becomes dominant. At temperatures over the Debye's temperature, the resistivity can be described as:

$$\rho = 1/\sigma = 1/[q^2 g(E_{\mathrm{F}})D] = 3/[q^2 g(E_{\mathrm{F}})v_{\mathrm{F}}^2 \tau_{\mathrm{F}}]. \tag{69}$$

By using Eq. (65) this expression can be rewritten as:

$$\rho = 3\sigma_{\mathrm{res}}kT/(q^2 v_{\mathrm{F}}) = \rho(T_0)(T/T_0), \tag{70}$$

where $T_0=300$ K, and $\rho(T_0) = 3\sigma_{\mathrm{res}}kT_0/(q^2 v_{\mathrm{F}})$, here $\sigma_{\mathrm{res}} = \sigma_{\mathrm{res}0}$ (Figure 10).

The resultant average relaxation time τ_{res} of RM electrons can be described as:

$$1/\tau_{\mathrm{res}} = (1/\tau_{\mathrm{eff}}) + (1/\tau_{\mathrm{def}}), \tag{71}$$

where

$$1/\tau_{\mathrm{eff}} = \sigma_{\mathrm{res}}N_{\mathrm{eff}}v_{\mathrm{F}} = \sigma_{\mathrm{res}}g(E_{\mathrm{F}})v_{\mathrm{F}}kT; \tag{72}$$

and

$$1/\tau_{\mathrm{def}} = \sigma_{\mathrm{def}}N_{\mathrm{def}}v_{\mathrm{F}}; \tag{73}$$

where σ_{def} is average scattering cross-section of the defects, N_{def} is the defect density.

Then the resistivity of many metals in temperature range from 1 K to 900 K can be described as:

$$\rho(T)=\rho_0 + \rho(T_0)(T/T_0)\eta(T/\Theta). \tag{74}$$

The calculated resistivity dependences on temperature in the range from 1 K to 900 K by Eq. (74) for elemental metals Al, Ag, Au, Sr, and W are presented in Figure 15 by solid lines with $T_0 = 300$ K. The Debye's temperature Θ is not completely constant (Schulze 1967; Kittel 1969), therefore it has been chosen for the best agreement to the experimental data; besides it can change with temperature. Here the calculations have been performed by using the constant values Θ which are presented in brackets near the metal chemical sign in Figure 15. It is seen that Eq. (74) well enough describes the experimental data for presented metals, though they have very different Fermi surfaces (Cracknell and Wong 1973). The resistivity dependence in the transition region can be described as T^{-n}, where n depends on the value ρ_0, and changes from 1 to 5.

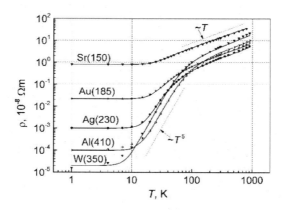

Figure 15. The resistivity of the group of elemental metals (Sr, Au, Ag, Al, and W) dependence on temperature: dots are experimental data (Knowledge door 2019); the solid lines are calculated by Eq. (74); in brackets near the metal chemical sign there is presented the Debye's temperature in K used for calculation.

The diffusion coefficient dependences on temperature for the same metals as in Figure 15 are presented in Figure 16. The diffusion coefficients have been calculated by Eq. (29) with DOS at the Fermi surface from the Figure 6, and by using the experimental resistivity values from Figure 15. A small diffusion coefficient of the RM electrons in gold at very low temperatures can be explained by large density of impurities and defects. Considering that the diffusion coefficient has been calculated by using only experimental data of the both resistivity and DOS from electronic heat capacity, so the obtained values of diffusion coefficients of the free RM electrons can be sustained as experimental ones.

Considering that the average Fermi velocity does not depend on temperature, it is simply to determine the RM electron mean free path as $l_F = 3D/v_F$, where v_F is evaluated by Eq. (67) (Figure 11). The RM electron mean free path for the same metals as in Figure 15 dependences on temperature are presented in the Figure 17. It seen that for W at $T<10$ K the RM electron mean free path exceeds 1 mm, while for Au it is only about 4 μm. The relaxation time of RM electrons at the Fermi surface is estimated as $\tau_F = l_F/v_F$. The relaxation times for elemental metals Ag, Al, Au Sr, and W dependences on temperature are shown in the Figure 18.

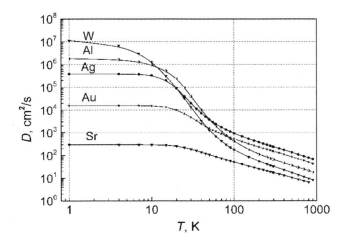

Figure 16. Diffusion coefficient [Eq. (29)] of the free RM electrons of elemental metals Ag, Al, Au, Sr, and W dependences on temperature.

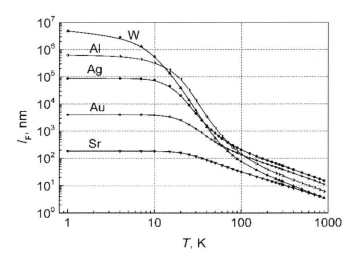

Figure 17. Mean free path of RM electrons of elemental metals Ag, Al, Au, Sr, and W dependences on temperature.

According to (Devillers 1984) in the linear resistivity dependence on temperature range (T>Θ), the relaxation time of the free electron can be estimated as $\tau_F \approx \hbar/kT$: at T=300 K $\tau_F \approx$26 fs. As it is seen from Figure 18, such relaxation time description can be used at T>Θ as an acceptable approximation. Using experimental data of the electrical resistivity dependence on temperature and applying the following calculation algorithm:

$$D(T) = 1/[q^2 g(E_F)\rho(T)] \Rightarrow l_F(T) = 3D(T)/v_F \Rightarrow \tau_F = l_F(T)/v_F, \tag{75}$$

one can find the basic RM electron transport characteristic dependences on temperature of any elemental metal in the normal (non-superconducting) state, because the DOS at Fermi surface for all elemental metals are known.

The presented results let somebody decide how to choose the needed metal thin films for cryogenic devices.

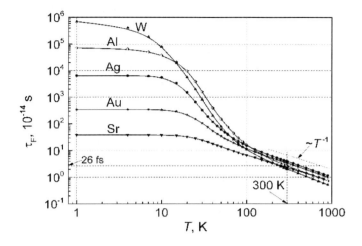

Figure 18. Relaxation time of RM electrons of elemental metals Ag, Al, Au, Sr, and W dependences on temperature; $\tau_F \approx 26$ fs at $T = 300$ K is according to (Devillers 1984).

The Other Applications of the Free RM Electrons Model

Plasma Frequency

The plasma frequency of metals is important to anyone who has an interest in such studies as photonics, plasmonics, infrared electronics and nanoparticle arrays. Usually the frequency-dependent (or a. c.) conductivity $\sigma(\omega)$ (Ashcroft and Mermin 1976; Sólyom 2009) is described as:

$$\sigma(\omega) = \sigma/(1 - j\omega\tau), \tag{76}$$

and the relative permittivity $\varepsilon_r(\omega)$ as:

$$\varepsilon_r(\omega) = 1 + j\sigma/[\varepsilon_0\omega(1 - j\omega\tau) = 1 - \omega_p^2[\omega(\omega + j/\tau)], \tag{77}$$

where $j = \sqrt{-1}$ is the imaginary unit, ε_0 is the electric constant, σ is the d. c. conductivity, τ is the average relaxation time of the free RM electrons, and:

$$\omega_p^2 = \sigma/(\varepsilon_0 \tau) \tag{78}$$

is the plasma frequency. The relative permittivity for $\omega\tau \gg 1$ is expressed as:

$$\varepsilon_r(\omega) = 1 - \omega_p^2/\omega^2. \tag{79}$$

When $\omega \geq \omega_p$, metal becomes transparent, and a radiation propagates through a metal. If somebody uses the Drude formula for conductivity, he gets the following expression for plasma frequency:

$$\omega_p^2 = q^2 n/(\varepsilon_0 m^*), \tag{80}$$

which can be applicable only for materials with non-degenerate electron gas. Here m^* is the effective mass of the electrons. Considering the basic Eq. (20) for d. c. conductivity, the plasma frequency for metals can be described as:

$$\omega_p^2 = q^2 g(E_F)v_F^2/(3\varepsilon_0). \tag{81}$$

It is seen that the plasma frequency for metals is determined by DOS and the Fermi velocity of free RM electrons at the Fermi surface, and it does not depend on the temperature and effective mass of the free RM electrons.

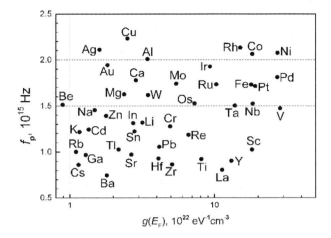

Figure 19. The plasma frequency [Eq. (81)] dependence on the DOS at the Fermi surface for elemental metals (in double logarithmic scale).

If somebody uses Eq. (80) for estimation of the plasma frequency of metals, for some of them he can get results close to the experimental one, but in general, he gets the wrong results. The plasma frequency $f_p = \omega_p/2\pi$ determined by Eq. (81) for elemental metals are shown in Figure 19.

According to the Sommerfeld model $\omega_p \sim \sqrt{n}$, and therefore there must be such proportionality $\omega_p \sim g^{3/2}(E_F)$, but plasma frequencies do not depend on the DOS at the Fermi surface: because the Fermi velocity in average is proportional to $1/\sqrt{g(E_F)}$ (Figure 20). As it is seen from Figure 19, the plasma frequency f_p is randomly distributed in the range from 0.75 THz (for Ba) to 2.23 THz (for Cu).

What properties of the free RM electrons cause the observed spread of plasma frequency data? In Figure 21, the plasma frequency and plasma energy dependence on the scattering cross-section of the free RM electrons by electronic defects are presented. These results directly show that the plasma frequency spreading is caused by particular electron scattering cross-section data due to electronic defects.

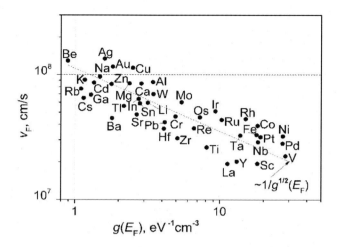

Figure 20. Fermi velocity of the free RM electrons dependence on the DOS at the Fermi surface.

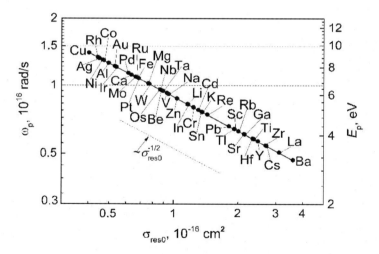

Figure 21. The plasma frequency ω_p (left scale) and plasma energy E_p (right scale) of the free RM electrons dependence on the scattering cross-section σ_{res0} of the electronic defects (in double logarithmic scale).

Magnetic Susceptibility of the Free Electrons of Metals

Considering the classical free electron theory, it has been believed that the paramagnetic susceptibility χ_{par} of the free electrons, due to their spin orientation parallel to applied magnetic field strength direction as for paramagnetic ions, must be in inverse proportion to temperature (Sólyom 2009; Ziman 1972):

$$\chi_{par} = n\mu_0 g_e^2 \mu_B^2/(4kT) \approx n\mu_0 \mu_B^2/(kT), \tag{82}$$

where μ_0 is the magnetic constant, μ_B is the Bohr magneton, $g_e \approx 2$ is the electronegativity of the electron. The measured magnetic susceptibility results show that it is temperature independent and much smaller. In derivation of Eq. (82), it has not been taken into account that free electrons in the conduction band which are well below the Fermi level energy cannot change their spin orientation due to the Pauli's exclusion principle. Considering that the effective free RM electron density is $n_{eff} = g(E_F)kT$, Eq. (82) can be rewritten as:

$$\chi_{par} = n_{eff}\mu_0 \mu_B^2/(kT) \approx \mu_0 \mu_B^2 g(E_F). \tag{83}$$

Therefore, this expression is well known as Pauli's paramagnetic susceptibility, and it can simply be explained on the basis of the free RM electrons, because only they can change their spin orientation in the magnetic field.

The diamagnetic susceptibility of the free electrons, due to their spatial (orbital) motion, creates the magnetic moment which is in the opposite direction to the applied magnetic field strength. The Landau diamagnetic susceptibility for free RM electrons is expressed as (Sólyom 2009; Ziman 1972):

$$\chi_{dia} = -(1/3)\chi_{par}, \tag{84}$$

and it partially compensates for the paramagnetic susceptibility. Thus, the magnetic susceptibility measurements cannot be used as acceptable estimation of the DOS at the Fermi surface, as they are obtained from the electronic heat capacity.

Kubo Formula and the Thermal Noise of the Free RM Electrons

The spectral density of the current fluctuations S_{i0} due to the free charge carrier random motion (thermal noise) at low frequencies $f \ll 1/(2\pi\tau)$ (here τ is the relaxation time of the free charge carriers) is described by Nyquist formula (Nyquist 1928; Van der Ziel 1986):

$$S_{i0} = 4kT(1/R) = 4kT\sigma(A/L), \tag{85}$$

where R is the resistance of tested sample, A and L are, respectively, the cross-section and the length of the sample. This expression is valid for all homogeneous materials at equilibrium state, and does not depend on the material nature. Eq. (85) also follows from the general Kubo formula for conductivity (Kubo 1965):

$$\sigma = (1/kT) \int_0^\infty < j(t + \tau_1) \cdot j(t) > d\tau_1, \tag{86}$$

i.e., the conductivity is unambiguously defined by the autocorrelation function on time

$$k_j(\tau_1) = < j(t + \tau_1) \cdot j(t) > \tag{87}$$

of the current density fluctuation $j(t)$ due to random motion of the charge carriers. The autocorrelation time for a steady state processes depends only on the time difference τ_1. The integral in Eq. (86), according to the Wiener-Khintchine theorem, is the spectral density of current density fluctuations (Van der Ziel 1986). It can explain, why many of the free RM

Study of the Transport of Charge Carriers in Materials ... 165

electron properties can be simply described by using characteristics of the random processes.

The current density fluctuations can be replaced by independent and RM charge carrier velocity fluctuations $j(t) = qv(t)$:

$$\sigma = (q^2/kT) \int_0^\infty k_v(\tau_1)d\tau_1 \tag{88}$$

where $k_v(\tau_1)$ is the autocorrelation function of the RM charge carrier velocity fluctuations. Considering the Wiener-Khintchine theorem at low frequencies, Eq. (88) can be expressed as:

$$\sigma = (q^2 n_{\text{eff}}/4kT)S_{v0} \tag{89}$$

where $S_{v0} = 4D$ is the spectral density of the velocity fluctuations (Palenskis et al. 2013), D is the diffusion coefficient of the RM charge carriers. Therefore, the conductivity is described by the same formula as in Eq. (20).

TRANSPORT OF CHARGE CARRIERS ON THE NORMAL STATE SUPERCONDUCTOR ON THE BASE OF YBCO

The general relations obtained earlier, based on the Fermi distribution of the free RM charge carriers, are applied for estimation of the transport characteristics of superconductors at temperatures well above the superconducting phase transition temperature T_c on the basis of the superconductor $YBa_2Cu_3Y_{7-x}$ (YBCO). Considering that for most cuprate superconductors for which doping has been optimized for maximum transition temperature T_c, the resistivity $\rho(T)$ is almost linear on T as for elemental metals (Anderson1997; Waldram 1996). Therefore, the conductivity has been interpreted by using the Drude model due to lattice scattering. The positive Hall coefficient shows that free charge carriers are holes in the superconductor YBCO. The Hall coefficient at temperatures higher than Debye's temperature for different metals has a negative or positive sign. The Hall coefficient measurement data of the normal state superconductor $YBa_2Cu_3Y_{7-x}$ with $x = 0$ in many cases are in inverse proportionality to temperature (Anderson1997; Dresselhaus 2001b; Waldram 1996). The charge carrier concentration n obtained from classical Hall coefficient expression $R_H = 1/qn$ varies proportional to T. It has been supposed that scattering rate $(1/\tau)$ is proportional to T^2. The infrared conductivity analysis (Waldram 1996) shows that the low frequency conductivity as function of T agree to d. c. conductivity, but the ratio n/m^* decreases very slightly with temperature, which does not agree with the Hall effect measurement results, and scattering rate $(1/\tau)$ from the infrared data is proportional to T rather than T^2, and plasma frequency $\omega_p^2 = \sigma/(\varepsilon_0 \tau) \approx$ const. Essentially, the constancy of the free carrier density has been established by the penetration depth measurements by muon spin rotation technique. In order to explain both the resistivity and the Hall coefficient dependences on temperature, there were proposed, two-band models (Eagles 1989; Jin and Ott 1988; Smith 1987; Xing, Liu and Ting 1988; Xing and Ting 1988) in terms of the total densities n and p, and the drift mobilities μ_n and μ_p, respectively, for electrons and holes:

$$\sigma = 1/\rho = qn\mu_n + qp\mu_p, \tag{90}$$

$$R_H = (p\mu_p^2 - n\mu_n^2)/[q(p\mu_p + n\mu_n)^2], \tag{91}$$

so taken that drift μ_n and μ_p and Hall mobilities for superconductor are the same due to that $\tau = \tau_F$. In this model the temperature dependence of the Hall coefficient reflects a temperature dependent compensation between electrons and holes. But it is not possible to uniquely determine n, p, μ_n and μ_p from data of $R_H(T)$ and $\rho(T)$. In order to explain $R_H(T) \sim 1/T$ and $\rho(T) \sim T$ behavior in terms of two-band model, it is necessary to assume very unusual particular relations among the characteristics. In the work (Eagles 1989), it was assumed that the densities of electrons and holes do not

Study of the Transport of Charge Carriers in Materials ...   167

change with temperature, but mobilities μ_n and μ_p vary in the following way:

$$\mu_p^{-1} = A(1 + BT), \tag{92}$$

$$\mu_n^{-1} = C(1 + DT), \tag{93}$$

where A, B, C, D are constants, and $B \neq D$. Eight parameters are adjusted in order to give an accurate fit to resistivity and Hall measurement data.

The properties of the high-T_c superconductors in the normal state have been overlooked by (Anderson 1997). He proposed that the key quantity is not the Hall coefficient but the Hall angle, which he describes as:

$$\Theta_H = \omega_c \tau_H = 1/(A + BT^2), \tag{94}$$

where A and B are constants; $\omega_c = qB_z/m^*$ is the cyclotron frequency, B_z is the magnetic flux density in z-direction. Thus, the relaxation time which determines the Hall angle must be proportional to T^{-2} rather than T^{-1}. From this model it follows that during measurement of resistivity well above T_c (without the magnetic field) the average relaxation time changes with temperature as $1/T$, but during the measurement of Hall effect in the presence of the weak magnetic field the average relaxation time must change as $1/T^2$. The measurement of the magnetoresistance in single-crystal $YBa_2Cu_3O_{7-\delta}$ above 95 K is smaller than 1 % of the resistance (Hikita and Suzuki 1989). Therefore, there is no real evidence that in a weak magnetic field the charge carrier relaxation time drastically changes. Ones also can see from Eqs. (51) and (52) that the same charge carrier relaxation time is in the conductivity expression and it gives that Hall coefficient does not depend on the relaxation time.

According to (Hirsch and Marsiglio 1991), the electrons and holes respond differently to variations of the temperature, and the sign of the Hall coefficient is determined by the competition of these two contributions, which is controlled by the doping level.

Considering that high-T_c superconductors are the materials with degenerate electron (and hole) gas, respectively, the electrons and holes cannot be described by classical model. So, Eq. (90) for conductivity and Eq. (91) for the Hall coefficient in two-band models (Eagles 1989; Jin and Ott 1988; Xing, Liu and Ting 1988; Xing and Ting 1988) have been derived on the misconstruction based that the drift and Hall mobilities for materials with degenerate charge carrier gases coincide, but such an assumption, as it was shown earlier, is completely incorrect. Besides, the conductivity of the material with degenerate charge carriers is determined by the effective densities of the RM charge carriers, but not by the total densities of electrons and holes.

The general expressions both of the conductivity and the Hall coefficient for materials with free RM charge carriers for two-band model is described (Bonch-Bruevitch and Kalashnikov 1990; Markiewicz 1988; Smith 1987; Ziman 1972) as:

$$\sigma = 1/\rho = \sigma_n + \sigma_p = qn\mu_n + qp\mu_p, \tag{95}$$

and

$$R_{H2} = (R_{Hp}\sigma_p^2 - R_{Hn}\sigma_n^2)/\sigma^2 = (\mu_{Hp}\sigma_p - \mu_{Hn}\sigma_n)/\sigma^2; \tag{96}$$

where according to (Palenskis 2015):

$$\sigma = qn_{eff}\mu_{ndrift} + qp_{eff}\mu_{pdrift}; \tag{97}$$

$$n_{eff} = g_n(E_F)kT; \tag{98}$$

$$p_{eff} = g_p(E_F)kT; \tag{99}$$

$$\mu_{Hp} = R_{Hp}\sigma_p = q\tau_F/m_0; \tag{100}$$

$$\mu_{Hn} = R_{Hn}\sigma_n = -q\tau_F/m_0; \tag{101}$$

$$\mu_{ndrift} = \mu_{pdrift} = qv_F^2\tau_F/(3kT) = qD/kT; \tag{102}$$

where $g_n(E_F)$ and $g_p(E_F)$, respectively, are the effective electron-like and hole-like DOS at the Fermi surface, and $g_{total}(E_F) = g_n(E_F) + g_p(E_F) =$ const is the total DOS at $E = E_F$.

Considering that electrons and holes randomly move with the Fermi velocity, their relaxation times at the Fermi surface are the same, so the cyclotron frequencies and absolute values of the Hall mobilities for separate charge carriers in the weak magnetic field are the same. The absolute values for one type of charge carriers over the T_c of the both Hall μ_{H1} and drift μ_{drift1} mobility dependence on temperature for single crystal of the YBCO are presented in Figure 22: for holes they have the positive signs, and for electrons they have negative signs. It is seen that absolute values of the Hall mobility of the RM charge carriers are proportional to T^{-1}, while the drift mobility change as T^{-2}, but in all cases the average relaxation time and the mean free path of the free charges decrease as T^{-1} with temperature increasing (Figure 23).

Inserting the Eqs. (98)–(102) into Eqs. (95) and (96), the conductivity and the Hall coefficient of two-band model for normal state superconductor can be expressed in the following form (Palenskis 2015):

$$\sigma = q^2v_F^2\tau_F[g_n(E_F) + g_p(E_F)]/3 = q^2v_F^2\tau_F g_{total}(E_F)/3 \tag{103}$$

$$R_{H2} = R_{H0}[1 - 2g_n(E_F)/g_{total}(E_F)], \tag{104}$$

where $R_{H0} = 3/[2qg_{total}(E_F)E_F]$ is the Hall coefficient for the case of the single type of the free RM charge carriers. From Eq. (103) it follows that relaxation time of the free RM charge carriers varies with temperature as $1/T$, and the same dependence on temperature is for charge carrier mean free path: $l_F = v_F\tau_F$ (Figure 24). Eq. (104) can be rewritten in more simple form:

$$R_{H2} = R_{H0}\eta, \tag{105}$$

the quantity η reflects the compensation of Hall voltages caused by holes and electrons in degenerate material. The quantity η can be named by Hall effect compensation factor. The Hall effect compensation factor η dependence on temperature is shown in Figure 23.

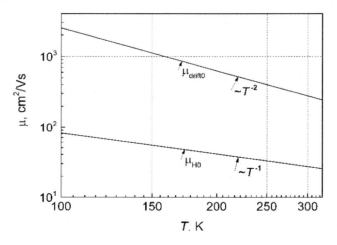

Figure 22. The absolute values of the Hall μ_{H0} and drift μ_{drift0} mobility dependences on temperature for one type of the RM charge carriers over the T_c for single crystal of the YBCO (for holes they have the positive signs, and for electrons they have negative signs).

Considering that R_{H0} is determined by Eq. (55), from the measurement of the Hall coefficient R_{H2}, one can determine only the compensation factor η, and so to determine the contributions of electron-like and hole-like DOS to the total one at Fermi surface, and they vary with temperature and other conditions. On the other hand, the Hall coefficient R_{H2} described as $1/(qn)$ has no relation to the total density n of the free charge carriers in the tested sample, and cannot be used for estimation of the n. Besides, the quantity R_{H2} is very dependent on oxygen content, pressure and imperfections of the crystal.

Study of the Transport of Charge Carriers in Materials ...

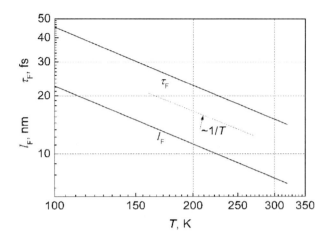

Figure 23. The average relaxation time τ_F (in fs) and the mean free path l_F (in nm) of the free RM charge carrier dependence on temperature over the T_c for single crystal of the YBCO.

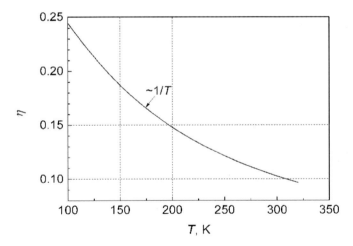

Figure 24. Hall effect compensation factor dependence on temperature for single crystal YBCO with maximum T_c.

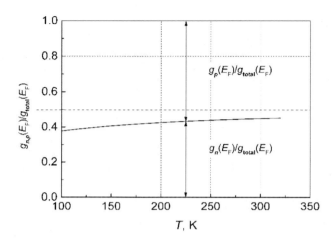

Figure 25. The redistribution of the electron-like and hole like DOS with temperature.

On the base of the Hall coefficient, Hal angle and Hall mobility measurements data for single crystals YBCO, whose doping has been optimized for maximum T_c, presented by (Carrington 1992; Eagles 1989; Forro 1989; Jin and Ott; Kostylev 1990), and in review works (Anderson 1997; Ginsberg 1990; Waldram 1996) at temperature range from 100 K to 300 K we can described by following relation:

$$\eta = R_{H2}(T)/R_{H0} = \mu_{H2}(T)/\mu_{H0}(T) = \cot\Theta_0(T)/\cot\Theta_2(T) \sim 1/T, \quad (106)$$

where index "$_0$" shows the parameters of the superconductor in the case of single type of the free RM charge carriers, and index "$_2$" – for two types (electrons and holes) of the free RM charge carriers.

The Hall effect compensation factor dependence on temperature is very sensitive to the superconductor doping level, DOS at the Fermi surface, imperfections, and crystal growth conditions. The electron-like and hole-like DOS at the Fermi surface dependences on temperature for single crystal YBCO is shown in Figure 25. When the electron-like and hole-like DOS at the Fermi surface are close one to other, the small changes of these densities with temperature can produce very significant changes in temperature dependence of the Hall coefficient and Hall mobility.

So, in order to explain the Hall effect properties, there is no need to account for the spinons, holons (Anderson 1997), spin bipolarons (Mott 1990) or other quasi-particles.

TRANSPORT OF ELECTRONS IN DONOR-DOPED SILICON AT ANY ELECTRON DENSITY

Silicon has been the main material for electronics for sixty years. Its basic characteristics have been intensively investigated theoretically and experimentally (Baker-Finch et al. 2014; Canali et al. 1975; Ingole et al. 2008; Jacoboni et al. 1997; Jain 1977; Kaiblinger-Grujin, Kosina and Selberherr 1997; Thobel, Sleiman and Fauquembergue 1997; Van Overstraten, DeMan and Mertens 1973; Xiao and Wei 1997). Transport phenomena play a fundamental role in solid-state devices, and the electronic technology requires an increase in our knowledge of transport quantities, such as the effective density of the RM charge carriers, drift mobility, diffusion coefficient as functions of different parameters, including the heavily doping.

The general expressions based on the Fermi distribution function of the free RM electrons there are applied for estimation of the base kinetic coefficients in donor-doped silicon at arbitrary degrees of the degeneracy of the electron gas under equilibrium conditions. The ratio of the classical value of the diffusion coefficient over the mobility D/μ for the majority and minority charge carriers satisfies the Einstein's relation kT/q (Jain 1977; Xiao and Wei 1997). In most of the published works, it is assumed that the Einstein's relation is valid only for materials with non-degenerate electron gas. Therefore, at thermal equilibrium for estimation of the relation between the diffusion coefficient and the mobility of the charge carriers in degenerate semiconductors, they have used the modified relation

(Ristič, S. D. 1979; Thobel, Sleiman and Fauquembergue 1997; Van Overstraten, DeMan and Mertens 1973):

$$D = n\mu/[q(\mathrm{d}n/\mathrm{d}E_\mathrm{F})], \tag{107}$$

where n is the total density of the free electrons in the conduction band, and E_F is the Fermi level energy; the similar relation is used for free holes in the valence band. Eq. (107) can be written as:

$$q^2 D(\mathrm{d}n/\mathrm{d}E_\mathrm{F}) = qn\mu, \tag{108}$$

where both sides mean the conductivity. But the left side

$$q^2 D(\mathrm{d}n/\mathrm{d}E_\mathrm{F}) = \int_0^\infty g(E)[\partial f(E)/\partial E_\mathrm{F}]\mathrm{d}E = q^2 D n_\mathrm{eff}/kT \tag{109}$$

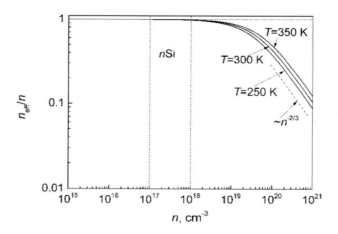

Figure 26. The ratio of the effective density of the free RM electrons to the total density of the free electrons n_eff/n dependence on the total density of electrons in the conduction band of nSi at three temperatures.

is proportional to the effective density of the free RM electrons at the Fermi surface, and is valid for homogeneous materials at any degeneration degree of the electron gas, while the right side is correct only for material with non-degenerate electron gas. Thus, Eq. (107) is not applicable for

Study of the Transport of Charge Carriers in Materials ... 175

description characteristic of semiconductor with a high density of the free charge carrier density. The ratio between the effective density n_{eff} of the RM electrons at the Fermi surface and the total density n of the free electrons in the conduction band of the nSi is described as:

$$n_{eff}/n = \int_0^\infty g(E)f(E)[1 - f(E)]dE / \int_0^\infty g(E)f(E)dE, \qquad (110)$$

and is shown in Figure 26 at three temperatures.

The results in Figure 26 well illustrate that the effective density of the RM electrons at the Fermi surface is equal to the total density of free electron in the conduction band of the nSi, when the shallow, completely ionized donor density is smaller than 10^{18} cm^{-3}, i.e., $n_{eff} = n$ at $n < 10^{18}$ cm^{-3}, and at n>10^{20} cm^{-3} n_{eff}/n decreases as $n^{-2/3}$. As it is seen from Eqs. (25) and (26), the diffusion coefficient and the drift mobility of the free RM electrons depend on their average kinetic energy <E>. The average kinetic energy for free RM electron is described as:

$$< E >= \int_0^\infty E g(E)f(E)[1 - f(E)]dE / n_{eff}, \qquad (111)$$

and its dependence on the total density of electrons in the conduction band of nSi at three temperatures are presented in Figure 27. There, at the same temperatures, the Fermi level energy change with the total density of the free electrons are also shown. From this figure it can be seen that for $n < 10^{18}$ cm^{-3} <E>=$(3/2)kT$, i. e. the classical statistics is applicable. Then the average kinetic energy of the free RM electron increases with n, and for n>10^{20} cm^{-3}, it coincides with the Fermi level energy.

From Eq. (20) it follows that such a general expression for the diffusion coefficient of the free RM electrons of nSi:

$$D = kT/[q^2\rho(T)n_{eff}] = kT\alpha_\varepsilon/[q^2\rho(T)n], \qquad (112)$$

where for $\alpha_\varepsilon = n/n_{eff} =< E >/(3kT/2)$ (from Eq. (26), and Figure 26). The obtained results at room temperature are presented in Figure 28.

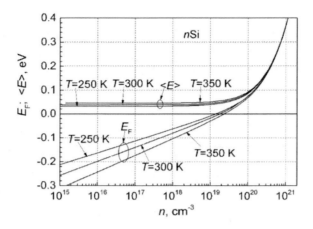

Figure 27. The Fermi level energy E_F and the average kinetic energy $<E>$ of the free RM electron dependence on the total density of electrons in the conduction band of nSi at three temperatures.

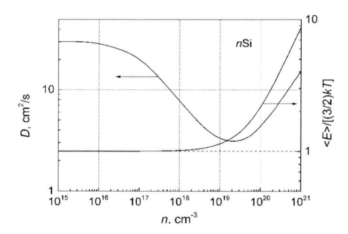

Figure 28. Diffusion coefficient D (left scale) and the ratio between the average kinetic and thermal energies $<E>/[(3/2)\,kT]$ (right scale) of the free RM electrons dependences on the total density of electrons in the conduction band of nSi at room temperature ($T = 300$ K).

There, the resistivity $\rho(T)$ dependence on the total free electron density in the conduction band at $T = 300$ K data are taken from references (Dargys and Kundrotas 1994; INSPEC 1998; Jacoboni et al. 1977; Sze and Kwok 2007). The calculated data of the quantity α_ε by Eq. (26) also are presented in Figure 28. The quantity α_ε at room temperature can be approximated by the simple expression: $\alpha_\varepsilon = (1 + 0.21 \cdot 10^{-19} n)^{0.7}$. It is well seen that diffusion coefficient increase at $n > 10^{19}$ cm^{-3} is caused by the increase of the average kinetic energy of the RM electrons.

Considering that Einstein's relation [Eq. (23)] between the diffusion coefficient and the drift mobility of the free RM electrons is valid at all free carrier densities in the conduction band, the drift mobility can be estimated in two ways:

$$\mu_{drift} = 1/(q\rho n_{eff}) = qD/kT; \tag{113}$$

here it can be pointed out that Einstein's relation is valid also for homogeneous materials with non-parabolic energy zones, because Einstein's relation shows that at equilibrium, the current components caused by drift and diffusion compensate one other, and the resultant total current is zero, and it does not depend on the total density of the free charge carriers.

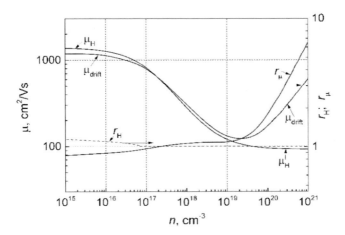

Figure 29. The Hall μ_H and drift μ_{drift} mobilities (left scale), Hall r_H and drift r_μ mobility factors (right scale) of the free RM electrons dependences on the total density of electrons in the conduction band of nSi at room temperature ($T=300$ K).

Comparison of the both drift and Hall mobility dependences on the total free carrier density in the conduction band of nSi at room temperature is presented in Figure 29. It is seen that at the degenerate electron gas (n $> 10^{19}$ cm^{-3}), the drift mobility exceeds by many times the Hall mobility, though in both cases the average relaxation time $<\tau> = \tau_F$. In Figure 29 you can also see the Hall factor r_H change from the scattering mechanism of the charge carriers in the investigated free carrier density range.

According to (Smith 1987), the Hall factor in this free carrier density interval can vary from $3\pi/4$ to 1. It is worthwhile to introduce the new factor: the drift mobility factor r_μ, which reflects the ratio between the drift and the Hall mobilities: μ_{drift}/μ_H. At $n<10^{18}$ cm^{-3} $r_\mu \approx 1/r_H$, at higher density n it is proportional to the average kinetic energy of the RM electrons.

CONCLUSION

The presented study has been written to analyze the charge carrier transport characteristics in materials with degenerate electron gas. It has been shown that erroneous values of parameters can be obtained, if the classical statistics are applied for estimation of the transport characteristics for such materials. The application of the Fermi-Dirac statistics and the stochastic description of the free RM charge carriers lead to presenting the general expressions for the conductivity, and the effective density of the free RM electrons, their diffusion coefficient and drift mobility, which are valid for all homogeneous materials with any degree of degeneracy of the electron gas. It was shown that Einstein's relation for one type of charge carriers (electrons or holes) in homogeneous materials is fundamental, and does not depend on the degree of degeneracy of the electron gas. It also does not depend on the non-parabolicity of the conduction band or on the scattering mechanism of charge carriers. By using the obtained general relations, we have estimated such parameters of the RM electrons: average diffusion coefficient, average drift mobility, mean free path, average kinetic velocity, average relaxation time and scattering cross-section at the Fermi surface, also Fermi energy and plasma frequency for 44 elemental metals at room temperature, by using only the experimental values of the resistivity and electronic heat capacity of these metals. It is shown that at temperatures above the Debye's temperature, the electron scattering cross-section in metals does not depend on temperature. The resistivity dependence on temperature in a very wide temperature range has been explained by electronic defects, accounting for the ratio of the exchange of thermal energies between the phonon and electron. There for the first time, for Ag, Al, Au, Sr, and W are presented the average diffusion coefficient, average drift mobility, mean free path, and average relaxation time of the RM electrons dependences on temperature from 1 K to 900 K.

On the basis of the effective density of the RM charge carriers, the general expressions for the Hall coefficient and the Hall mobility for two types of the charge carriers (electrons and hole) have been obtained, and applied to the explanation of the Hall coefficient of superconductor $YBa_2Cu_3O_7$ dependence on temperature in the normal state. It is shown that from the Hall effect, measurement data in degenerate material with electrons and holes, it is only possible to determine the compensation degree factor, and the parts of the electron-like and hole-like DOS at the Fermi surface, but not the total charge carrier density.

The obtained general relations based on the Fermi distribution of the free electrons are applied for estimation of their diffusion coefficient and drift mobility in donor-doped silicon at an arbitrary degree of degeneracy of the electron gas under equilibrium conditions. It is shown that drift mobility of the RM electrons at high doping levels considerably exceeds the Hall mobility, and that the Einstein's relation between the diffusion coefficient and the drift mobility of the RM one type charge carries (electrons or holes) is conserved at any level of degeneracy.

REFERENCES

Abrikosov, A. A. 1988. *Fundamentals of the Theory of Metals.* Amsterdam: North-Holland Publication.

Alloul, H. 2011. *Introduction to the Physics of Electrons in Solids.* Berlin Heidelberg: Springer-Verlag.

Anderson, P. W. 1997. *The Theory of Superconductivity in the High-Tc Cuprates. Part I: Ch. 3 Normal State Properties in the High-Tc Superconductors: Evidence for Non-Fermi Liquid States.* Princeton: Princeton University Press

Ashcroft, N. W., N. D. Mermin. 1976. *Solid State Physics.* Chapter 2. New York: Harcourt College Publishers.

Baker-Finch, S. C., McIntosh, K. R., Yan, D., Fong, K. Ch., and Kho, T. C. 2014. "Near-Infrared Free Carrier Absorption in Heavily Doped Silicon." *Journal of Applied Physics* 116: 063106-1(12). http://dx.doi.org/10.1063/1.4893176.

Bisquert, J. 2008. "Interpretation of Electron Diffusion Coefficient in Organic and Inorganic Semiconductors with Broad Distributions of States." *Physical Chemistry and Chemical Physics* 10: 3175–3194. http://dx.doi.org/10.1039/b719943k.

Blakemore, J. S. 1985. *Solid State Physics.* Cambridge: Cambridge University Press.

Bonch-Bruevitch, V. L., and Kalashnikov S. G. 1990. Physics of Semiconductors (in Russian). Moscow: Nauka.

Canali, C., Majni, G., Minder, R., and Ottaviani. G. 1975. "Electron and Hole Drift Velocity Measurements in Silicon and Their Empirical relation to Electric Field and Temperature." *IEEE Transactions on Electron Devices* ED-22: 1045-1047. https://doi.org/10.1109/T-ED.1975.18267.

Carrington, A., Makenzie, A. P., Lin, C. T., and Cooper, J. R. 1992. "Temperature dependence of the Hall angle in $YBa_2(Cu_{1-x}Co_x)_3$ $O_{7-\delta}$." Physical Review Letters 69: 2855-2857.

Chopra, K. L. 1979. *Thin Film Phenomena.* Malabar: R. E. Krieger Publication Company, Science.

Cracknell, A. P., and Wong, K. C. 1973. *The Fermi Surfaces: Its Concept, Determination, and Use in the Physics of Metals.* Oxford: Clarendon Press.

Dargys, A., and Kundrotas, J. 1994. *Handbook on Physical Properties of* Ge, Si, GaAs *and* InP. Vilnius: Science and Encyclopedia Publishers.

Devillers, M. A. C. 1984. "Lifetime of Electrons in Metals at Room Temperature." *Solid State Communications* 49: 1019–1022. https://doi.org/10.1016/0038-1098(84)90413-7.

Dresselhaus, M. S. 2001a. *Solid State Physics. Part I: Transport Properties of Solids.* 6.732, Fall, downloaded as PDF.

Dresselhaus, M. S. 2001b. *Solid State Physics. Part IV: Superconducting Properties of Solids.* 6.732, Fall, downloaded as PDF.

Dugdale, J. S. 2010. *The Electrical Properties of Disordered Metals.* Cambridge: Cambridge University Press. http://dx.doi.org/10.1017/CBO09780511629020.

Eagles, D. M. 1988. "Concentrations and Mobilities of Holes and Electrons in a Crystal of a 90 K Oxide Y-Ba-Cu-O." *Solid State Communications* 69: 229–234. http://dx.doi.org/10.1016/0038-1098(89)90840-5.

Forro, L., Raki, M., Henry, J. Y., and Ayache, C. "Hall Effect and Thermoelectric Power of an $YBa_2Cu_3O_{6.8}$ Single Crystal." Solid State Communications 69: 1097-1101. http://dx.doi.org/10.1016/0038-1098(89)90493-6.

Gall, D. 2016. "Electron Mean Free Path in Elemental Metals." *Journal of Applied Physics* 119: 085101−1(5). https://doi.org/10.1063/1.4942216.

Ginsberg, D. M. ed. 1990. *Physical Properties of High Temperature Superconductors.* University of Illinois, Urbana-Champaign: World Scientific Publishing Co, Pte, Ltd. https://doi.org/10.1142/1023.

Hirsch, J. E., and Marsiglio, F. 1991. "Hole Superconductivity in Oxides: A Two-Band Model." *Physical Review B* 43: 424-434. https://doi.org/10.1103/PhysRevB.43.424.

Hikita, M., and Suzuki, M. 1989. "Magnetoresistance and Thermodynamic Fluctuations in Single-Crystal $YBa_2Cu_3O_y$." *Physical Review B* 39: 4756–4759.
http://doi.org/10.1103/PhysRevB.39.4756.

Ingole, S., Manandhar, P., Chikkannanavar, S. B., Akhadov, E. A., and Picraux, S. T. 2008. "Charge Transport Characteristics in Boron-Doped Silicon Nanowires." *IEEE Transactions on Electron Devices* 55:2931–2938.
http://doi.org/10.1109/TED.2008.2005175.

INSPEC. 1998. *Properties of Silicon, EMIS Data Series No. 4. INSPEC.* London: The Institute of Electrical Engineering.

Jacoboni, C. Canali, C., Ottaviani, G, Quaranta, A. A. 1997. "A Review of Some Transport Properties of Silicon." *Solid-State Electronics* 20: 77-89.
https://doi.org/10.1016/0038-1101(77)90054-5.

Jain, R. K. 1977. "Calculation of the Fermi Level, Minority Carrier Concentration, Effective Intrinsic Concentration, and Einstein Relation in *n*- and *p*-Type Germanium and Silicon." *Physica Status Solidi* 42: 221-226.
http://dx.doi.org/10.1002/pssa.2210420123.

Jin, R., and Ott, H. R. 1988. "Hall Effect of $YBa_2Cu_3O_{7-\delta}$ Single Crystals." *Physical Review B* 57: 13872–13877.
http://dx.doi.org/10.1103/PhysRevB.57.13872.

Kaiblinger-Grujin, G., Kosina, H., and Selberherr, S. 1997. "Monte Carlo Simulation of Electron Transport in Doped Silicon." *IEEE Xplore*: 444-449.
http://dx.doi.org/10.1109/HPC.1997.592188.

Kaveh, M., and Wiser, N. 1984. "Electron-Electron Scattering in Conducting Materials." *Advances in Physics* 33: 257–372.

Kittel, Ch. 1969. *Thermal Physics.* New York: John Wiley and Sons, Inc.

Kittel, Ch. 1976. *Introduction to Solid State Physics.* New York: John Wiley and Sons, Inc.

Kostylev, V. A., Chebotayev, N. M., Naumov, S. V., and Samochvalov, A. A. 1990. "Kinetic Properties of Single Crystals $YBa_2Cu_3O_x$ (x=6.95

and 6.1)." *Superconductivity: Physics, Chemistry, Engineering* 3(1): 156.

Lide, D. R. 2003-2004. *Handbook of Chemistry and Physics*, 84 Edition. Florida: CRC Press.

Linde, J. O. 1964. "The Effective Mass of the Conduction Electrons in Metals and the Theory of Superconductivity." *Physics Letters* 11 :199–201.

Lundstrom, M. 2014. *Fundamentals of Carrier Transport*. Cambridge: Cambridge University Press.

Markiewicz, K. R. 1988. "Simple Model for the Hall Effect in $YBa_2Cu_3O_{7-\delta}$." *Physical Review B* 38:5010–5011. http://dx.doi.org/10.1103/PhysRevB.38.5010.

Mott, N. F. 1990. "Hall effect above T_c in the superconductor $YBa_2Cu_3O_{7-\delta}$ (YBCO)." *Philosophical Magazine Letters* 62: 273-275.

Nyquist, H. 1928. "Thermal Agitation of Electric Charge in Conductors." *Physical Review* 32: 110-113.

Palenskis, V. 1990. "Flicker Noise Problem (review, in Russian)." *Lithuanian Journal of Physics* 30: 107–152.

Palenskis, V. 2013. "Drift Mobility, Diffusion Coefficient of Randomly Moving Charge Carriers in Metals and Other Materials with Degenerate Electron Gas." *World Journal of Condensed Matter Physics* 3: 73–81. http://dx.doi:104236/wjcmp.2013.31013.

Palenskis, V. 2014a. "The Effective Density of Randomly Moving Electrons and Related Characteristics of Materials with Degenerate Electron Gas." *AIP Advances* 4: 047119–1(9). http://dx.doi.org/10.1063/1.4871757.

Palenskis, V. 2014b. "Transport of Electrons in Donor-Doped Silicon at Any Degree of Degeneracy of Electron Gas." World Journal of Condensed Matter Physics." 4: 123–133. http://dx.doi:104236/wjcmp.2014.43017.

Palenskis, V. 2015. "Transport Characteristics of Charge Carriers in Normal State Superconductor $YBa_2Cu_3O_{7-\delta}$." *World Journal of Condensed Matter Physics* 5: 118–128. http://dx.doi:104236/wjcmp.2015.53104.

Palenskis, V., A. Juškevičius, and A. Laucius. 1985. "Mobility of Charge Carriers in Degenerate Materials." *Lithuanian Journal of Physics* 25: 125–132.

Palenskis, V., Pralgauskaitė, S., Maknys, K., and Matulionis, A. 2013. "Thermal Noise and Drift Mobility of Randomly Moving Electrons in Homogeneous Materials with Highly Degenerate Electron Gas." Paper presented at International Conference on Noise and Fluctuations, IEEE Proc. 22nd ICNF2013, 4 p., Montpellier, June 23–28.

Palenskis, V., and Žitkevičius, E. 2018. "Phonon Mediated Electron-Electron Scattering in Metals." *World Journal of Condensed Matter Physics* 8: 115–129. http://dx.doi:104236/wjcmp.2018.83008.

Palenskis, V., and Žitkevičius, E. 2020. "Analysis of Transport Properties of the Randomly Moving Electrons in Metals." *Materials Science* 26: No. 2 (*accepted for publication*).

Ristič, S. D. 1979. "An Approximation of the Einstein Relation for Heavily Doped Silicon." Physica Status Solidi (a) 52: K129–K132. http://dx.doi.org/10.1002/pssa.2210520250.

Rossiter, P. L. 2014. *The Electrical Resistivity of Metals and Alloys.* Cambridge: Cambridge University Press.

Schulze, G. E. R. 1967. *Metallphysik.* Berlin: Akademie-Verlag.

Seeger, K. 2004. *Semiconductor Physics: An Introduction.* Berlin: Springer-Verlag.

Smith, R. A. 1987. *Semiconductors.* 2nd Edition. Cambridge: Cambridge University Press.

Sólyom, J. 2009. *Fundamentals of the Physics of Solids.* Berlin: Springer. http://doi:10.1007/978-3-642-04518-9.

Sondheimer, E. H. 2001. "The Mean Free Path of Electrons in Metals." *Advances in Physics* 50: 499–537.

Sommerfeld, A., and Bethe, H. 1967. *Elektronentheorie der Metalle.* Berlin: Springer-Verlag.

Sze, S. M., and Kwok, K. Ng 2007. *Physics of Semiconductor Devices.* New Jersey: John Wiley & Sons, Inc.

Thobel. J. L., Sleiman, A., and Fauquembergue, R. 1997. "Determination of Diffusion Coefficients in Degenerate Electron Gas Using Monte Carlo Simulation." *Journal of Applied Physics* 82: 1220–1226. http://dx.doi.org:10.1063/1.365892.

Van der Ziel, A. 1986. *Noise in Solid State Devices and Circuits.* New York: John Wiley & Sons.

Van Overstraeten, R. J., DeMan, H. J., and Mertens, R. P. 1973. "Transport Equations in Heavily Doped Silicon." *IEEE Transaction on Electron Devices* ED-20:290-298. http://dx.doi.org:10.1109/T-ED.1973.17642.

Waldram, J. R. 1996. *Superconductivity of Metals and Cuprates. Ch. 14. Normal-State Transport Properties in the Cuprates.* London: The Institute of Physics.

Wert, Ch., and Thomson, R. M. 1964. *Physics of Solids.* New York: McGraw-Hill Book Company.

Wilson, A. H. 1958. *The Theory of Metals.* Cambridge: Cambridge University Press.

Xiao, Z.-X., and Wei, T.-L. 1997. "Modification of Einstein Equation of Majority- and Minority-Carriers with Band Gap Narrowing Effect in n-Type Degenerate Silicon with Degenerate Approximation and with Non-Parabolic Energy Bands." IEEE Transaction on Electron Devices 44:913-914. http://dx.doi.org:10.1109/16.568061.

Xing, D. Y., Liu, M., and Ting, C. S. 1988. "Out-off-Plane Transport Mechanism in the High-T_c Oxide Y-Ba-Cu-O." *Physical Review B* 38: 11992–11995. http://dx.doi.org/10.1103/PhysRevB.38.11992.

Xing, D. Y., and Ting, C. S. 1988. "Two-Band Model for Anisotropic Hall Effect in High-T_c Oxide Y-Ba-Cu-O." *Physical Review B* 38: 5134–5137.

http://dx.doi.org/10.1103/PhysRevB.38.5134.

Ziman, J. M. 1972. *Principles of the Theory of Solids*. Cambridge: Cambridge University Press.
http://dx.doi.org/10.1017/CBO9781139644075

Ziman, J. M. 2001. *Electrons and Phonons. The Theory of Transport Phenomena in Solids, Oxford Classic Texts in the Physical Sciences*. Oxford: Oxford University Press.

Knowledgedoor.com. Last entered May 2019.
http://www.knowledgedoor.com/2/elements_handbook/electrical_resistivity.html.

In: Electron Gas: An Overview
Editor: Tata Antonia

ISBN: 978-1-53616-428-2
© 2019 Nova Science Publishers, Inc.

Chapter 3

ENHANCED OUTPUT POWER OF INGAN-BASED LIGHT-EMITTING DIODES WITH AlGaN/GaN TWO-DIMENSIONAL ELECTRON GAS STRUCTURE

Jae-Hoon Lee[*]
Yield Enhancement Team, Foundry,
Samsung Electronics Co., Ltd,
Giheung, Korea

ABSTRACT

We demonstrate high-performance InGaN-based light-emitting diodes (LEDs) with tunneling-junction-induced 2-D electron gas (2DEG) at an AlGaN/GaN heterostructure, which is inserted in the middle of the P^{++}-GaN contact layer of a conventional LED structure. The output power of a LED with a 2DEG insertion layer shows 20% enhancement compared to that of a conventional LED at 350 mA. This enhancement in output power for the LED with a 2DEG insertion layer could be attributed

---

[*] Corresponding Author's E-mail: jaehoon03.lee@samsung.com.

to both enhanced hole-injection efficiency and lateral current spreading by the presence of 2DEG at the AlGaN/GaN heterostructure.

Keywords: AlGaN/GaN, heterostructure, InGaN, LED, 2DEG (two-dimensional electron gas), tunneling junction

1. INTRODUCTION

Group III–nitride semiconductors and their ternary solid solutions are very promising as the candidates for both short wavelength optoelectronics and power electronic devices [1-4]. The AlGaN/GaN heterostructure field effect transistors (HFETs) have a great potential for future high-frequency and high-power applications because of the intrinsic advantages of materials such as wide bandgap, high breakdown voltage, and high electron peak velocity (electron mobility in excess of 2000 cm^2/Vs and peak velocity approaching 3×10^7 cm/s at room temperature) [5]. Major developments in wide-gap III-nitride semiconductors have led to commercial production of high-brightness light emitting diodes (LEDs). The InGaN-based LEDs have already been extensively used in full-color displays, traffic displays, and other various applications such as projectors, automobile headlights, and general lightings.

In particular, white LEDs based on InGaN/GaN quantum well heterostructure are regarded as the most promising solid-state lighting devices to replace conventional incandescent or fluorescent light. In spite of their recent success, however, the output power of InGaN/GaN LEDs needs to be further improved. There have been intensive researches for the improvement of light extraction efficiency and the enhancement of brightness in the LED by using surface roughening, patterned sapphire substrate (PSS), flip-chip bonding, laser lift-off, and photonic crystal structure [6-11].

Although the technology has matured, there still exist challenges in growing a highly p-type doped layer and in optimizing the related device fabrication process. The lateral LEDs use lateral carrier injection, which usually suffers from non-uniform current spreading due to high p-GaN resistivity [12]. The current crowding effect causes not only a local light emission, but also strongly affects the reliability of the device. It is, therefore, essential that a highly resistive p-GaN contact layer must be combined with transparent metal surface coverage for better current spreading in lateral LEDs. Various techniques for achieving better current spreading and for improving the light extraction efficiency from the LEDs have been developed [13, 14]. The tunneling junction is shown to be an effective method to reduce the operation voltage. Recently, the n^+/p^+-GaN tunnel junctions have been applied to obtain the uniform luminescence and exhibited an improved radiative efficiency [15].

J. Sheu et al. reported that the n^+-InGaN/GaN short period superlattices (SPS) tunneling contact layer instead of high-resistivity p-GaN as a top contact layer could be a viable solution [16]. Low operation voltage GaN-based LED, which utilizes an Mg-doped AlGaN/GaN strained layer superlattice (SLS) contact layer, has been also demonstrated [17]. The p-type AlGaN/GaN SLS can increase the ionization efficiency of acceptors due to two-dimensional hole gas (2DHG), which results in much smaller lateral resistivity in the in-plane lateral direction.

However, the application of the 2DHG can be limited because the mobility of 2DHG is much lower than that of two-dimensional electron gas (2DEG) formed at the n-type AlGaN/GaN hetrostructure [18]. In this work, the n-AlGaN/GaN tunnel junction which is inserted in the middle of the p^{++}-GaN contact layer, was proposed for the first time to improve the hole-injection efficiency and the lateral current spreading, accomplished by optimizing the thickness and the doping concentration for the AlGaN/GaN heterostructure.

2. EXPERIMENTS

2.1. Growth of LED Structure with AlGaN/GaN Spreading Layer

The LED sample proposed in this work was grown on (0001) cone-shape-patterned-sapphire substrates using metal-organic chemical vapor deposition (MOCVD) [19]. The designed layer structure for the LED with total thickness of about 6 μm consists of undoped GaN/Si-doped n-GaN, five pairs of InGaN-GaN multiple quantum wells (MQWs), Mg-doped p-AlGaN/GaN SLS electron blocking layer (EBL), Mg-doped 60 nm-thick p^{++}-GaN layer, tunnel junction current spreading layer which consists of 27 nm-thick Si-doped AlGaN layer on about 5 nm-thick undoped-GaN layer, and finally Mg-doped 10 nm-thick p^{++}-GaN top contact layer as shown in Figure 1 (a). For comparison, two additional LED samples, which do not have 2DEG current spreading layer, were also grown; the one was grown with only the 10 nm-thick undoped GaN layer without growing the 25 nm-thick Si-doped AlGaN layer and the other was grown without both layers (corresponding to a conventional LED structure). However, the actual thicknesses of bottom p^{++}-GaN layer, undoped-GaN layer, Si-doped AlGaN layer, and top p^{++}-GaN contact layer were slightly different, confirmed by transmission electron microscope (TEM) measurement as shown in the inset of the Figure 1 (a). Prior to growing the AlGaN/GaN heterostructure, the growth was interrupted for 3 min. to lower the growth pressure. The reason for lowering the pressure is because a higher quality of AlGaN layer is usually grown at a lower pressure. The Mg-doping concentration both in top contact and bottom p^{++}-GaN layer and the Si-doping concentration in AlGaN layer were around 1×1020 and 1×1018 cm-3, respectively, which are confirmed by the secondary ion mass spectroscope (SIMS) measurement as shown in Figure 1 (b). Al content in AlGaN layer is 20% determined by high resolution x-ray diffraction (HRXRD).

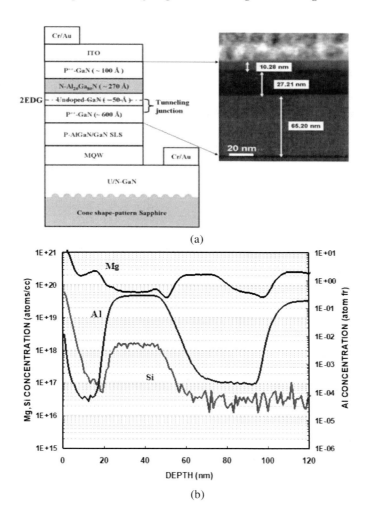

Figure 1. (a) Schematic view and TEM image, and (b) SIMS profile of the LED structure with AlGaN/GaN current spreading layer.

2.2. Fabrication of LED with AlGaN/GaN spreading layer

Devices of 260 μm × 670 μm and 1000 μm × 1000 μm dimensions were fabricated by a normal LED chip process using ITO transmittance for p-contact and Cr/Au metals for n-contact, respectively. After that, the sapphire substrate was thinned to about 80 μm thickness by using backside lapping and polishing [20]. The samples were mounted on sideview pakage and SUNNIX 5 high power package, respectively. The LED chips were mounted on lead frames and connected by Au wires. The current–voltage (I-V) characteristics were measured at room temperature using an HP4156 and Tektronix 370 semiconductor analyzer. The output power of the fabricated LED was measured by using integrating sphere to collect the light emitted in all directions from the LEDs at room temperature.

3. RESULTS AND DISCUSSION

3.1. Characteristics of AlGaN/GaN Spreading Layer

To measure the 2DEG mobility and a sheet carrier concentration of the proposed AlGaN/GaN structure, a control sample was prepared on the LED structure without growing the top p^{++}-GaN contact layer In general, a 2DEG mobility and a sheet carrier concentration for a typical AlGaN/GaN HFET structure, are known to be about 1000 ~ 2000 cm^2/Vs and 4×10^{12} ~ 1×10^{13}/cm^2, respectively. However, the 2DEG mobility for the 27 nm-thick AlGaN/GaN structure investigated in this work, grown on the p^{++}-GaN layer of the LED sample, is decreased to 320 cm^2/Vs. The sheet carrier concentration, instead, was greatly increased to about 7×10^{13}/cm^2, which is an order higher than that of a normal AlGaN/GaN heterostructure. The carrier type of AlGaN/GaN/p^{++}-GaN heterostructure is n-type, confirmed by Hall measurement, even though the Mg doping in n-AlGaN/GaN layer is higher than Si doping as shown in the SIMS data. This is because the activation efficiency of the Mg atom in AlGaN/GaN layer is much lower [21, 22] than that of the Si atom and thus the electron concentration is higher than hole concentration so that the structure exhibits n-type conduction. Figure 2 (a) and (b) show atomic force microscopy (AFM) data for 3 μm-thick semi-insulting GaN template for normal HFET structure and 6 μm-thick LED structure prior to the growth of tunnel junction, respectively. The surface roughness of the LED structure is 0.8 nm, much rougher than the corresponding value of 0.3 nm for the semi-insulating GaN layer. The 2DEG mobility can be decreased due to interface scattering because the interface becomes much rougher when the heterostructure is grown on the p-AlGaN/GaN SLS or the MQW active region than when it is grown on normal undoped GaN layer as shown in Figure 2. The 2DEG mobility also can be decreased due to impurity scattering caused by the unintentionally introduced Mg atoms after growing the p-type layer [23, 24].

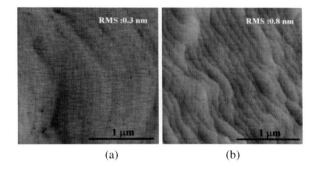

Figure 2. AFM images for (a) 3 μm-thickness semi-insulting GaN buffer and (b) 6 μm-thickness LED structure without AlGaN layer.

On the other hand, the sheet carrier density can be increased by the existence of not only the surface state charges on the AlGaN surface, but also the residual impurities in the AlGaN layer and all the layers underneath the heterostructure [25]. If the total integrated charges are donor-like, then the sheet carrier density must be increased to satisfy the charge neutrality [18, 26]. Compared to typical values of mobility of >~1500 cm^2/Vs and sheet carrier density of ~ 10^{13}/cm^2 for normal AlGaN/undoped HFET structure, the AlGaN/GaN heterostructure grown on the LED exhibits very low mobility of 320 cm^2/Vs and very high sheet carrier density of 7 × 10^{13}/cm^2. This is because the 2DEG mobility is greatly affected by the scattering mechanisms and the sheet carrier density is increased by the increased total donor-like charges in the structure, as discussed above. In addition, another possible explanation for such a high sheet carrier density observed in the 2DEG-LED structure is that the growth of the 2DEG-LED structure (with thickness of ~ 6 µm) may induce a strong tensile stress due to its complicated layer structure such as AlGaN/GaN tunnel junction, thick n-GaN, InGaN/GaN MQW, and p-AlGaN/GaN EBL layer. This tensile stress would increase the polarization-induced 2DEG density and also other tensile stress related charges which increases the sheet carrier density in the structure [27-29]. The structure with this extremely high 2DEG density is important because it plays a role of current spreading layer, even though the experimental observance of such a high 2DEG density is not clearly understood yet.

3.2. I-V Characteristics of Side-view LED with AlGaN/GaN Spreading Layer

I-V characteristics of all fabricated side-view LEDs are shown in Figure 3. The forward voltage of LEDs at 20 mA were approximately 3.20, 3.24, and 3.28 V for the LED without AlGaN/GaN current spreading layer (Ref-LED; conventional LED), the LED with undoped GaN interlayer only (GaN-LED), and the LED with AlGaN/GaN current spreading layer (2DEG-LED), respectively. The forward voltage of the 2DEG-LED is a little higher than that of the Ref-LED. This is because the inserted n-AlGaN/GaN current spreading layer increases the series resistance of the device which acts as a voltage drop as shown in inset of Figure 2. However, the leakage currents of the Ref-, the GaN-, and the 2DEG-LED were -18, -20, and -17 nA at a reverse voltage of -10 V respectively, showing negligible difference between samples, which demonstrates that the insertion of the current spreading layer does not degrade the electrical characteristics of the LEDs.

Enhanced Output Power of Ingan-Based Light-Emitting Diodes ... 195

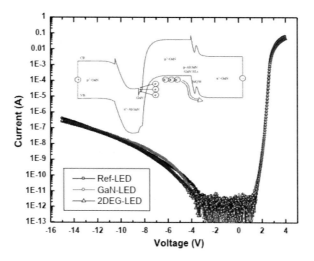

Figure 3. The I-V characteristics of Ref-LED, GaN-LED and 2DEG-LED. The inset shows a schematics view of band structure for 2DEG-LED at forward bias.

3.3 Optical Characteristics of Side-view LED with AlGaN/GaN Spreading Layer

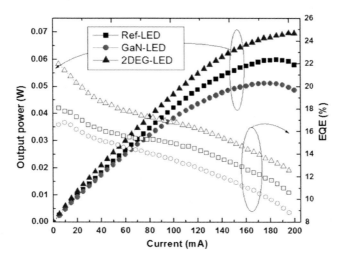

Figure 4. The light output power and external quantum efficiency of Ref-LED, GaN-LED, and 2DEG-LED as a function of operation current.

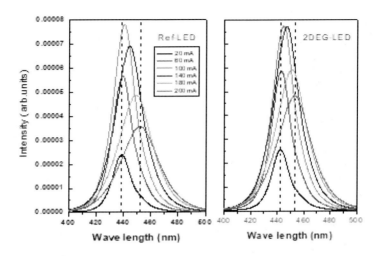

Figure 5. The electron luminescence spectrum of Ref-LED and 2DEG-LED as a function of operation current.

Figure 4 shows the light output power as well as external quantum efficiency (EQE) as a function of operation current. The total output power of the devices was measured by using 0.6t side-view package with a horizontal thickness of 6 mm without phosphor to collect the light emitted in all directions from the LEDs. The output power of the Ref-, GaN-, and 2DEG-LED at 20 mA was measured as 9.7, 9.2, and 11.4 mW (with corresponding EQE of 17.3, 16.2, and 20.3%), respectively, showing ~ 17% improvement in output power for the 2DEG-LED compared to the conventional Ref-LED. The improvement in output power for the 2DEG-LED is because the lateral distribution of the injected carriers becomes very uniform by the presence of AlGaN/GaN current spreading layer. As shown in inset of Figure 3, tunneling junction is formed between the quantum well with high 2DEG density, formed at AlGaN/GaN hetero-interface, and the bottom P^{++}-GaN layer [15, 16]. With applying a forward bias on the LED, the electrons in valance band of the bottom p^{++}-GaN layer are supposed to tunnel into the empty sub-bands of the 2DEG valley at AlGaN/GaN heterostructure, which in turn increases the hole concentration in the bottom p^{++}-GaN layer and hence increases the hole injection efficiency into the active MQW region. On the other hand, the light output power of the GaN-LED at 20 mA is slightly decreased, 5% lower than that of the conventional Ref-LED, because the 10 nm-thick inserted undoped GaN layer alone does not play a role of current spreading layer, but rather increases the series resistance of the p^{++}-GaN layer. The improvement in output power for the 2DEG-LED is further enhanced at higher current level, as shown in Figure 4. For instance, at 200 mA, the output power of 2DEG-LED is 20% higher than that of Ref-LED. We believe that the improvement in output power is because the high 2DEG density formed at AlGaN/GaN interface effectively spreads the current and also well dissipates the heat generated during the high current operation. However, the output power of GaN-LED is 21% smaller than that of Ref-LED. This is because the joule heating becomes very significant when high current flows through the device with increased series resistance caused by inserting the undoped GaN layer. The less joule-heating in the 2DEG-LED was also confirmed from the electron luminescence (EL) spectrum shift measurement with increasing injection currents as shown in Figure 5. The dashed lines in the figure indicate the amount of the red-shift. The EL peak positions for both the Ref-LED and the 2DEG-LED were red-shifted [11]. However, the amount of the shift is quite less for the 2DEG-LED as the current level increases from 20 to 200 mA; red-shift from 438 to 452 nm for the Ref-LED and from 443 to 453 nm for the 2DEG-LED.

3.4. I-V Characteristics of Power LED with AlGaN/GaN Spreading Layer

The I-V characteristics of the fabricated large chip LEDs are shown in Figure 6. The forward voltages of LEDs at 20 mA are approximately 2.85 V and 2.95 V for the LED without an n-AlGaN/GaN current spreading layer (Ref-LED) and the LED with an n-AlGaN/GaN current spreading layer (2DEG-LED), respectively. The slightly high forward voltage for the 2DEG LED is because the resistance of the device is a little higher at low current level with the existence of the additional AlGaN/GaN layer. As the current increases to 350 mA, the forward voltage becomes almost same; 3.39 V for Ref-LED and 3.42 V for the 2DEG-LED. As the current further increases to 1,500 mA, however, the forward voltage of the 2DEG-LED becomes lower than that of the Ref-LED (5.14V for Ref-LED and 4.87 V for 2DEG-LED), which indicates that extremely high density-2DEG formed at AlGaN/GaN layer effectively spreads the current at high current level. The leakage currents of the Ref-LED and 2DEG-LED at a reverse voltage of -20 V were around -103 and -32μA, respectively, as shown in the inset in Figure 6. The presence of AlGaN/GaN current spreading layer at least does not result in deteriorating I-V characteristics.

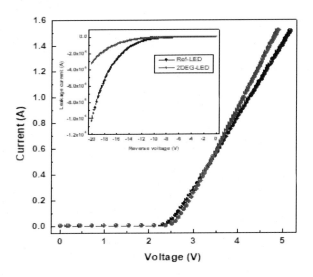

Figure 6. I-V characteristics of Ref-LED and 2DEG-LED. The inset shows a leakage current.

3.5. Simulation of AlGaN/GaN Spreading Layer

Figure 7 shows the simulated band diagram of the n-AlGaN/GaN heterostructure inserted in the middle of the p^{++}-GaN contact layer at equilibrium and under forward bias using Silvaco simulation program.

Enhanced Output Power of Ingan-Based Light-Emitting Diodes ...

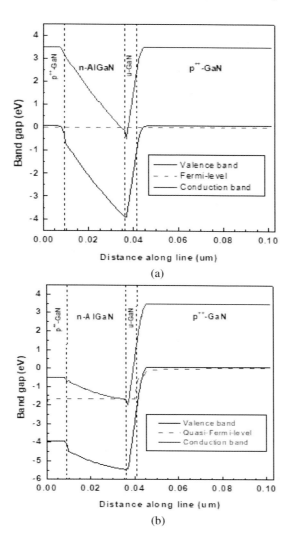

Figure 7. (a) Band diagram of the tunnel junction at equilibrium. (b) Band diagram of the tunnel junction with applied voltage V = 4V.

Tensile stress is not included in the band diagram to avoid complication, but only the basic physics is used to draw the band diagram. The effect of unintentionally doped Mg on drawing the band diagram is not important because the activation efficiency of the Mg atom is very low and hence the Si-doped AlGaN layer still exhibits n-type, as discussed above, even though the Mg doping shown in SIMS data is very high. For the simulation, thickness of 10, 27, 5, and 60 nm were used for top p^{++}-GaN, n-$Al_{0.2}Ga_{0.8}N$, undoped-GaN, and bottom p^{++}-GaN, respectively. Other values for p-, n-type doping density, and for the 2DEG sheet carrier density were 1×10^{20} cm^{-3}, and 1×10^{18} cm^{-3}, and 7×10^{13} cm^{-2}, respectively. As shown in Fig 7 (a), the conduction band edge near AlGaN/GaN interface is far below the Fermi energy which means that the band structure is highly degenerated and hence the 2DEG density in the quantum well, formed at the interface between AlGaN and GaN, is very high. The interface layer with this high 2DEG density serves as an induced n^{++}-layer and forms the n^{++}/p^{++} tunneling junction with the bottom p^{++}-GaN layer, similar to the conventional tunnel junction formed at the heavily junction formed at the heavily doped n^{++}/p^{++}-GaN junction [30, 31]. When a forward voltage is applied, the band diagram changes as if the tunneling junction becomes reversed biased as shown in Figure 7 (b). The valence band of the bottom p^{++}-GaN layer is raised above the conduction band of the AlGaN/GaN layer, which increases the tunneling probability of electrons from the valence band of the p^{++}-GaN layer into the empty sub-bands of the 2DEG valley at n-AlGaN/GaN heterostructure, leaving holes in the bottom p^{++}-GaN layer. At low current level, the measured forward voltage of 2DEG-LED was a little higher than that of the Ref-LED as observed above. This is because the inserted AlGaN/GaN layer itself and two heterointerfaces with finite band discontinuity formed at the top and the bottom p^{++}-GaN layer act as a voltage drop as shown in Figure 7. However, the band discontinuity has a crucial advantage of confining the tunneled electrons from the valence band of the p^{++}-GaN layer to the AlGaN/GaN heterinterface, which results in larger electron density at the interface. This is different from the case for the conventional n^{++}/p^{++}-GaN junction, where the tunneled electrons quickly moves toward the Ohmic contact. The electron confinement leads to smaller resistivity in the in-plane lateral direction and hence the current spreads easily in the lateral directions, which reduces the current crowding in lateral type LED device. At higher current level, the observed forward voltage of the 2DEG-LED was lower than that of the Ref-LED. This is because the 2DEG confined at n-AlGaN/GaN interface the empty sub-bands of the 2DEG valley at AlGaN/GaN heterostructure as mentioned before, which in turn increases the hole concentration in the bottom p^{++}-GaN layer and hence increases effectively spreads the current during the high current operation.

3.6. Optical Characteristics of Power LED with AlGaN/GaN Spreading Layer

Figure 8 shows the light output power of Ref-LED and 2DEG-LED as a function of operation current. The total output power of the devices was measured by using high power package without phosphor to collect the light emitted in all directions from the LEDs. The output power of the Ref- and 2DEG-LED at 350 mA was measured as 99 and 121 mW, respectively, showing ~20% improvement in output power for the 2DEG-LED compared to the Ref-LED. The improvement in output power for the 2DEG-LED is because the concentration of the injected carriers may be large and uniform as well by the presence of n-AlGaN/GaN current spreading layer [17].

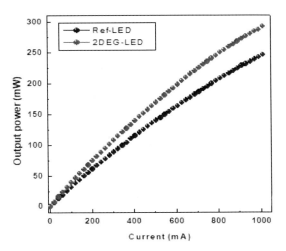

Figure 8. The light output of Ref-LED and 2DEG-LED as a function of operation current. The inset shows a schematics view of band structure for 2DEG-LED at reverse bias.

With applying a forward bias on the LED, the electrons in the valance band of the bottom p^{++}-GaN layer are supposed to tunnel into the hole injection efficiency into the active MQW region. Moreover, the improvement in output power for the 2DEG-LED is further enhanced at higher current level, as shown in the figure. The output power of 2DEG-LED is 25% higher than that of Ref-LED at 800 mA. This large enhancement in output power for the LED with 2DEG-insertion layer could be attributed to the enhanced hole injection efficiency into the active MQW region along with the lateral current spreading by the presence of 2DEG at n-AlGaN/GaN heterostructure.

3.7. Dynamic Resistance and I-V Characteristics of Power LED with AlGaN/GaN Spreading Layer

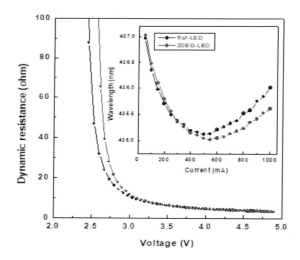

Figure 9. Dynamic resistances of Ref-LED and 2DEG-LED as a function of operation voltage. The inset shows a EL spectrum wavelength of Ref-LED and 2DEG-LED as a function of operation current.

Figure 9 shows the dynamic resistances for both the Ref-LED and the 2DEG-LED as a function of operation voltage. When the forward bias of 2.6 V is applied, the 2DEG-LED exhibits about two times larger resistance than the Ref-LED. As described above, n-AlGaN/GaN heterostructure interface of the 2DEG-LED act as a voltage drop layer in the vertical direction of current flow at low bias. However, when the forward bias increases to above 3.5V, the dynamic resistance of 2DEG-LED becomes smaller than that of Ref-LED. This originates from the fact that the injected 2DEG electrons from the tunneling junction are well confined at n-AlGaN/GaN heterostructure and hence effectively spread the device current in the lateral direction of the device. This is the reason why the dynamic resistance of the 2DEG-LED is smaller than that of the Ref-LED at high bias operation. The less joule-heating is also expected for the 2DEG-LED because of the smaller dynamic resistance for the 2DEG-LED, which is confirmed from the electroluminescence (EL) spectrum shift measurement as shown in the inset of Figure 9. The EL peak positions for both the Ref-LED and the 2DEG-LED were red-shifted [32]. However, the amount of the shift is quite less for the 2DEG-LED as the current level increases from 350 to 1000 mA; red-shift from 435.2 to 436.1 nm for the Ref-LED and from 435.3 to 435.6 nm for the 2DEG-LED.

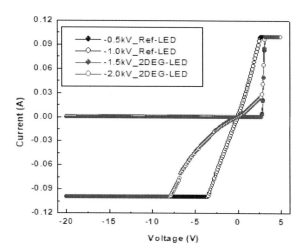

Figure 10. I-V characteristics of Ref-LED and 2DEG-LED after negative voltage human body model ESD stress.

GaN LEDs are known to fail at a relatively high positive-voltage electrostatic discharge (ESD) stress and also at low negative-voltage ESD stress [33, 34]. Figure 10 shows the I-V characteristics for LEDs after negative voltage human body model ESD stress. A reverse leakage current of 2 μA at -5 V was used as a failure criterion after ESD stress test. The Ref-LED was failed after applying a negative ESD voltage higher than 500 V. On the other hand, the failure was not observed for the 2DEG-LED even after applying a negative 1500 V ESD voltage. In other words, InGaN–based LEDs with tunneling junction-induced 2DEG at n-AlGaN/GaN heterostructure is potentially useful in fabricating a reliable nitride-based LED. This improved ESD voltage for the 2DEG-LED is probably due to the annealing effect during the growth interruption. The growth interruption introduced to lower the pressure, as described before, may also play a role of an intermediate temperate annealing, which suppresses propagation of threading dislocation in the grown and results in the improved ESD voltage for the 2DEG-LED [35].

CONCLUSION

We have investigated the electrical and optical properties of the LED with 2 DEG current spreading layer formed by tunneling of electrons at n-AlGaN/GaN heterostructure. The 2DEG-LED exhibited 20% higher output power at 350 mA compared to that of the conventional Ref-LED. The improvement was more prominent at high current level operation because the high 2DEG density is very effective both in spreading the current in lateral direction and in dissipating the heat generated in the device during the high current operation.

REFERENCES

[1]   Nakamura, S., M. Senoh, S. Nagahama, N. Iwasa, T. Yamada, T. Mat-sushita, Y. Sugimoto, and H. Kiyoku, *Jpn. J. Appl. Phys.*, vol. 36, pp. L1059-L1061, 1997.

[2] Nakamura, S., M. Senoh, N. Iwasa, and S. Nagahama, *Jpn. J. Appl. Phys.*, vol. 34, pp. L797-L799, 1995.

[3] Lee, J. H., D. Y. Lee, B. W. Oh, and J. H. Lee, *IEEE Trans. Electron Devices*, vol. 57, no. 1, pp. 157-163, 2010.

[4] Youn, D. H., J. H. Lee, V. Kumar, K. S. Lee, J. H. Lee, and I. Adesida, *IEEE Trans. Electron Devices*, vol. 51, no. 5, pp. 785-789, 2004.

[5] Smorchkova, P., S. Keller, S. Heikman, C. R. Elsass, B. Heying, P. Fini, J. S. Speck, and U. K. Mishra, *Appl. Phys. Lett.* vol. 77, pp. 3998-4000, 2000.

[6] Hsieh, M. Y., C. Y. Wang, L. Y. Chen, T. P. Lin, M. Y. Ke, Y. W. Cheng, Y. C. Yu, C. P. Chen, D. M. Yeh, C. F. Lu, C. F. Huang, C. C. Yang, and J. J. Huang, *IEEE Electron Device Lett.*, vol. 29, pp. 658-660, July, 2008.

[7] Wang, S. J., K. M. Uang, S. L. Chen, Y. C. Yang, S. C. Chang, T. M. Chen and C. H. Chen, *Appl. Phys. Lett.*, vol. 87, pp. 011111-011113, 2005.

[8] Lee, J. H., J. T. Oh, S. B. Choi, Y. C. Kim, H. I. Cho and J. H. Lee, *IEEE Photon. Technol. Lett.*, vol. 20, pp. 345 2008.

[9] Huang, S. H., R. H. Horng, K. S. Wen, Y. F. Lin, K. W. Yen, and D. S. Wuu, *IEEE Photon. Technol. Lett.*, vol. 18, no.24, pp. 2623 – 2625, 2006.

[10] Wuu, D. S., W. K. Wang, K. S. Wen, S. C. Huang, S. H. Lin, R. H. Horng, Y. S. Yu and M. H. Pan. *J. Electrochem. Soc.*, vol. 153, pp. G765, 2006.

[11] Huang, H. W., C. H. Lin, Z. K. Huang, K. Y. Lee, C. C. Yu, H. C. Kuo, *IEEE Electron Device Lett.*, vol. 30, pp. 1152-1154, Nov., 2009.

[12] Guo X. and E. F. Schubert, *J. Appl. Phys.*, vol. 90, no. 8, pp. 4191-4195, Oct. 2001.

[13] Jang, J. S. *Appl. Phys. Lett.*, vol. 93, no. 8, pp. 081118-081120, Aug. 2008.

[14] Liu, Y. J., C. H. Yen, L. Y. Chen, T. H. Tsai, T. Y. Tsai, and W. C. Liu, *IEEE Electron Device Lett.*, vol. 30, no. 11, pp. 1149-1151, Nov. 2009.

[15] Jeon, S. R., Y. H. Song, H. J. Jang, G. M. Yang, S. W. Hwang, and S. J. Son, *Appl. Phys. Lett.*, vol. 78, pp. 3265-3267, 2001.

[16] Sheu, J. K., J. M. Tsai, S. C. Shei, W. C. Lai, T. C. Wen, C. H. Kou, Y. K. Su, S. J. Chang, and G. C. Chi, *IEEE Electron Device Lett.*, vol. 22, no. 10, pp. 460-462, Oct. 2001.

[17] J. K. Sheu, G. C. Chi, and M. J. Jou, *IEEE Electron Device Lett.*, vol. 22, pp. 160-162, Apr. 2001.

[18] O. Ambacher, J. Smart, J. R. Shealy, N. G. Weimann, K. Chu, M. Murphy, W. J. Schaff, L. F. Eastman, R. Dimitrov, L. Wittmer, M. Stutzmann, W. Rieger, and J. Hilsenbeck, *J. Appl. Phys.* vol. 85, pp. 3222-2123, 1999.

[19] Lee, J. H., J. T. Oh, Y. C. Kim and J. H. Lee, *IEEE Photon. Technol. Lett.*, vol. 20, pp. 1563-1565, 2008.

[20] Lee, J. H., N. S. Kim, D. Y. Lee and J. H. Lee, *IEEE Photon. Technol. Lett.*, vol. 21, pp. 1151-1153, 2009.

[21] Tanaka, T., A. Watanabe, H. Amano, Y. Kobayashi, I. Akasaki, S. Yamazaki, and M. Koike, *Appl. Phys. Lett.*, vol. 65, pp. 593 - 594, 1994.

[22] Liu, Y. J., T. Y. Tsai, C. H. Yen, L. Y. Chen, T. H. Tsai, and W. C. Liu, *IEEE J. Quantum Electron.*, vol. 46, no. 4, pp.492-498, 2010.

[23] Keller, S., G. Parish, P. T. Fini, S. Heikman, C. H. Chen, N. Zhang, S. P. DenBasars, and U. K. Mishra, *J. Appl. Phys.*, vol. 86, pp.5850–5857, Nov. 1999.

[24] Lee, J. H., J. H. Kim, S. B. Bae, K. S. Lee, J. S. Lee, J. W. Kim, S. H. Hahm, and J. H. Lee, *Phys. Status Solidi C*, vol. 0, pp. 240-243, 2002.

[25] Im, K. S., J. B. Ha, K. W. Kim, J. S. Lee, D. S. Kim, S. H. Hahm, J. H. Lee, *IEEE Electron Device Lett.*, vol. 31, pp. 192-194, 2010.

[26] Ibbetson, J. P., P. T. Fini, K. D. Ness, S. P. DenBaars, J. S. Speck, and U. K. Mishra, *Appl. Phys. Lett.*, vol. 77, pp. 250, 2000.

[27] Dems M. and W. Nakwaski, *Semicond. Sci. Technol.*, vol. 18, pp. 733, 2003.

[28] Paskova, T. L. Becker, T. Bottcher, D. Hommel, P. Paskov and B. Monemar, *J. Appl. Phys.*, vol. 102, pp. 123507, 2007.

[29] Brunner, F., A. Knauer, T. Schenk, M. Weyers and T. Zettler, *J. Crystal Growth*, vol. 310, pp. 2432, 2008.

[30] BenDaniel D. J. and C. B. Duke, *Phys. Rev.*, vol. 152, pp. 683–692, 1966.

[31] Boucart, J. C., Starck, F. Gaborit, A. Plais, N. Bouche, E. Deroui, J. C. Remy, J. Bonnet-Gamard, L. Goldstein, C. Fortin, D. Carpentier, P. Salet, F. Brillouet and J. Jacquet, *IEEE J. Sel. Topics Quantum Electron*, vol. 5, pp. 520, 1999.

[32] Huang, S. Y., R. H. Horng, P. L. Liu, J. Y. Wu, H. W. Wu, and D. S. Wuu, *IEEE Photon. Technol. Lett.*, vol. 20, pp.797-799. 2008.

[33] Shei, S. C., J. K. Sheu and C. F. Shen, *IEEE Electron Device Lett.*, vol. 28, pp. 346-348, 2007.

[34] S. J. Chang, C. H. Chen, Y. K. Su, J. K. Sheu, W. C. Lai, J. M. Tsai, C. H. Liu and S. C. Chen, *IEEE Electron Device Lett.*, vol. 24, pp. 129 2003.

[35] Bourret-Courchesne, E. D., S. Kellermann, K. M. Yu, M. Benamana, Z. Liliental-Weber and J. Washburn, *Appl. Phys. Lett.*, vol. 77, no. 22, pp. 3562-3564, 2000.

BIOGRAPHICAL SKETCH

Jae-Hoon Lee

Affiliation: Samsung Electronics Co., Ltd

Education: Ph.D. degrees in electronic engineering from Kyungpook National University

Business Address: Yield Enhancement Team, Foundry, Samsung Electronics Co., Ltd, Giheung, 17113, Korea

Research and Professional Experience: GaN power device, LED, CMOS

Professional Appointments: Principal engineer

Honors: Dr. Lee is profiled in Marquis who's who in the world 2010, Great Minds of the 21st Century (ABI), Top 100 Engineers 2010 (IBC), and 2000 Outstanding Intellectuals of the 21st Century (IBC) in 2010.

Publications from the Last 3 Years:

1. JH Lee, KS Im, JK Kim, JH Lee, "Performance of Recessed Anode AlGaN/GaN Schottky Barrier Diode Passivated With High-Temperature Atomic Layer-Deposited Al2O3 Layer," *IEEE Transactions on Electron Devices* 66 (1), 324-329, 2019.
2. KS Im, JH Seo, S Vodapally, IM Kang, JH Lee, S Cristoloveanu, JH Lee, "Performance enhancement of AlGaN/GaN nanochannel omega-FinFET," *Solid-State Electronics* 129, 196-199, 2017.

INDEX

#

2DEG (two-dimensional electron gas), ix, 5, 11, 12, 15, 16, 20, 26, 33, 42, 47, 52, 102, 110, 114, 187, 188, 189, 190, 192, 193, 194, 195, 196, 197, 198, 200, 201, 202, 203, 204

A

accounting, 8, 148, 179
additives, 86, 111
Aharonov-Bohm effect, viii, 2
AlGaN/GaN, v, vii, ix, 187, 188, 189, 190, 191, 192, 193, 194, 195, 196, 197, 198, 200, 201, 202, 203, 204, 208
amplitude, 5, 11, 15, 22, 27, 30, 34, 38, 41, 58, 73, 74, 75, 87, 91, 103, 111, 113, 114
annihilation, 45, 57, 148
atoms, 3, 4, 142, 145, 146, 149, 193

average kinetic energy, 135, 146, 175, 176, 177, 178

B

bandgap, 188
base, 71, 114, 172, 173
beryllium, 133
bias, 140, 195, 196, 198, 201, 202
Boltzmann constant, 35
Boltzmann distribution, 15, 27
bonding, 188
breakdown, 188

C

candidates, 188
carbon, 2, 3, 5, 6, 49, 85, 112, 114, 115
carbon nanotubes, 2, 3, 112, 114, 115
charge density, 146, 147
chemical, 4, 11, 12, 13, 15, 16, 21, 22, 23, 26, 27, 28, 34, 35, 36, 39, 40, 42, 54, 111, 126, 134, 155, 156, 190
chemical potential, 4, 11, 12, 13, 15, 16, 21, 22, 23, 26, 27, 28, 34, 35, 36, 39, 40, 42, 54, 111, 126, 134
chemical vapor deposition, 190
chirality, 104, 105, 106, 107

210 Index

classical statistics, ix, 124, 125, 130, 135, 175, 178

compensation, 166, 170, 171, 172, 179

conduction, 3, 13, 18, 75, 114, 125, 126, 127, 128, 130, 131, 133, 134, 142, 145, 163, 174, 175, 176, 177, 178, 192, 200

conduction band, 114, 125, 126, 127, 128, 130, 131, 133, 134, 142, 145, 163, 174, 175, 176, 177, 178, 200

conductivity, v, vii, viii, ix, 1, 2, 4, 6, 7, 43, 45, 46, 47, 49, 52, 53, 55, 56, 57, 58, 61, 62, 66, 68, 70, 71, 73, 74, 75, 76, 78, 79, 80, 81, 82, 83, 84, 85, 86, 87, 111, 112, 113, 124, 125, 127, 131, 132, 133, 134, 135, 136, 137, 139, 140, 142, 143, 144, 154, 159, 160, 164, 165, 167, 168, 169, 174, 178

conservation, 59, 98, 114

contour, 3, 9, 15, 18, 19

Coulomb interaction, 3

crystal structure, 188

cyclotron angular frequency, 139, 140

D

damping, viii, 2, 7, 62, 68, 80, 81, 84, 93, 99, 100, 101, 107, 108, 109, 110, 112, 113, 114

Debye's temperature, 138, 145, 146, 148, 149, 153, 155, 156, 166, 179

defects, 146, 147, 149, 155, 156, 157, 161, 162, 179

degenerate electron gas, vii, viii, 2, 5, 7, 22, 24, 25, 43, 49, 77, 80, 81, 98, 105, 107, 109, 110, 111, 123, 124, 125, 130, 134, 135, 140, 141, 160, 173, 174, 178

density of electrons, 43, 56, 76, 84, 86, 98, 103, 174, 175, 176, 177

density of the free electrons, 128, 129, 142, 174, 175

density-of-states, 125, 126

diffusion, vii, ix, 124, 134, 136, 137, 138, 139, 141, 143, 149, 150, 151, 152, 153, 154, 157, 165, 173, 175, 177, 178, 179

diffusion coefficient, vii, ix, 124, 134, 136, 137, 138, 139, 141, 143, 149, 150, 151, 152, 153, 154, 157, 165, 173, 175, 176, 177, 178, 179

dispersion, 5, 47, 49, 59, 61, 74, 77, 79, 80, 81, 84, 85, 86, 89, 93, 98, 100, 104, 109, 112

distribution, 11, 17, 29, 104, 125, 126, 127, 128, 131, 133, 139, 146, 147, 151, 165, 173, 179, 196

distribution function, 11, 104, 125, 126, 127, 173

doping, 165, 167, 172, 173, 179, 189, 190, 192, 200

drift mobility, ix, 124, 125, 134, 135, 137, 138, 140, 141, 143, 169, 173, 175, 177, 178, 179

dynamic conductivity, 2, 6, 7

E

effective density, vii, viii, 123, 124, 125, 127, 128, 129, 130, 131, 132, 133, 134, 136, 138, 143, 149, 173, 174, 175, 178, 179

effective density of randomly moving electrons, vii, 124, 125

effective density of the free RM electrons, 125, 174, 178

Einstein's relation, 134, 139, 173, 177, 178, 179

electric field, 45, 56, 75, 125, 139

electrical conductivity, 125, 127, 133, 134, 135

electroluminescence, 203

electromagnetic, viii, 2, 6, 47, 55, 59, 60, 61, 62, 68, 74, 75, 78, 80, 81, 85, 112, 113

Index 211

electromagnetic waves, viii, 2, 6, 47, 61, 62, 68, 74, 80, 81, 85, 112, 113

electron diffusion coefficient, 124, 150

electron drift mobility, 124

electron scattering, vii, ix, 4, 124, 148, 149, 150, 155, 161, 179

electron spin waves, viii, 2, 7, 110

electronic defects, 146, 147, 149, 155, 161, 162, 179

electronic heat capacity, 130, 131, 139, 147, 157, 164, 179

electronic thermal conductivity, 132, 136

emission, 189

energy, vii, viii, ix, 1, 2, 4, 6, 7, 8, 12, 13, 15, 16, 18, 22, 23, 24, 27, 28, 29, 30, 31, 32, 33, 34, 35, 36, 37, 41, 42, 43, 44, 48, 50, 51, 52, 57, 58, 59, 60, 68, 69, 71, 75, 78, 84, 87, 91, 92, 94, 98, 99, 100, 102, 103, 104, 109, 110, 111, 112, 113, 124,125, 126, 127, 128, 129, 130, 131, 133, 135, 138, 140, 142, 143, 145, 146, 147, 148, 152, 154, 155, 161, 162, 163, 174, 175, 176, 177, 178, 179, 200

engineering, 207

entropy, 4, 14, 26, 42

equilibrium, 134, 143, 164, 173, 177, 179, 198, 199

excitation, 29, 148

F

fabrication, 189

Fermi (level) energy, ix, 22, 35, 37, 43, 48, 51, 52, 58, 59, 60, 69, 71, 78, 84, 87, 94, 99, 100, 110, 113, 124, 125, 126, 127, 128, 129, 130, 131, 138, 142, 143, 147, 152, 154, 163, 174, 175, 176, 179, 200

Fermi distribution (function), 11, 104, 125, 126, 127, 139, 165, 173, 179

Fermi velocity, 51, 86, 102, 133, 140, 143, 149, 152, 153, 157, 160, 161, 162, 169

Fermi-Dirac statistics, ix, 124, 125, 145, 178

field theory, 119

fluctuations, 146, 148, 164, 165

formula, viii, 2, 9, 11, 12, 13, 14, 16, 19, 21, 29, 31, 34, 36, 39, 45, 51, 54, 56, 68, 70, 71, 74, 75, 76, 80, 82, 84, 92, 93, 94, 95, 96, 97, 98, 99, 131, 160, 164, 165

free RM electrons, 125, 137, 144, 146, 147, 149, 150, 151, 152, 153, 154, 157, 160, 161, 162, 163, 173, 174, 175, 176, 177, 178

freedom, 17, 29, 41, 42

G

geometry, 3, 6, 112

Germany, 117, 120

graph, viii, 2, 100

graphene sheet, 6

growth, 99, 190, 192, 194, 204

growth pressure, 190

H

Hall angle, 139, 167, 180

Hall coefficient, 139, 140, 141, 142, 143, 144, 145, 165, 166, 167, 168, 169, 170, 172, 173, 179

Hall effect, 124, 139, 141, 142, 145, 166, 167, 170, 171, 172, 173, 179, 183

Hall effect of metals, 124

Hall mobility, ix, 124, 138, 140, 141, 144, 169, 172, 173, 178, 179

Hamiltonian, 3

Hartree-Fock, 4, 103

heat capacity, viii, 2, 4, 5, 13, 14, 17, 24, 25, 26, 29, 34, 35, 36, 37, 38, 41, 42, 43, 111, 125, 130, 131, 139, 147, 157, 164, 179

helicity, 68, 101

212 Index

heterostructure, vii, ix, 6, 187, 188, 189, 190, 192, 193, 196, 198, 200, 202, 203, 204

hexagonal lattice, 152

human body, 203, 204

I

ideal, 4, 141, 146, 147

impurities, 146, 155, 157, 193

inequality, 13, 42, 50, 54, 59, 80, 105

InGaN, vii, ix, 187, 188, 189, 190, 194, 204

insulators, 125

integral probability distribution function, 128, 129

integration, 3, 9, 11, 14, 15, 19, 34, 69, 93, 94

interface, 193, 196, 200, 203

ons, 146, 147, 163

K

kinetics, 3, 115, 121, 122

L

lead, vii, 1, 131, 178, 192

light, vii, ix, 128, 129, 187, 188, 192, 195, 196, 201

light-emitting diodes, vii, ix, 187, 188

low temperatures, 17, 36, 146, 157

luminescence, 189, 196, 197

M

magnetic field, vii, viii, 1, 2, 3, 5, 6, 7, 14, 18, 20, 27, 28, 30, 33, 36, 37, 40, 41, 42, 43, 47, 49, 50, 55, 56, 59, 74, 75, 83, 84, 91, 92, 99, 102, 103, 104, 109, 110, 111, 112, 139, 140, 163, 167, 169

magnetic field effect, 5

magnetic moment, 18, 26, 30, 44, 103, 163

magnetization, 4, 5, 14, 26, 114

magnetoresistance, 167

mass, 4, 7, 8, 14, 17, 18, 30, 31, 38, 39, 42, 43, 44, 56, 75, 86, 91, 103, 110, 112, 113, 114, 125, 131, 140, 141, 142, 152, 160, 190

materials, viii, 123, 124, 125, 126, 130, 133, 134, 135, 141, 160, 164, 168, 173, 174, 177, 178, 188

mean free path, vii, ix, 124, 145, 147, 149, 150, 157, 158, 169, 171, 179

measurement, 37, 112, 131, 139, 142, 145, 154, 155, 166, 167, 170, 179, 190, 192, 197, 203

measurements, 140, 164, 166, 172

metals, vii, viii, 4, 123, 124, 125, 128, 130, 131, 133, 134, 135, 136, 137, 138, 140, 141, 142, 143, 145, 146, 148, 149, 150, 151, 152, 153, 154, 156, 157, 158, 159, 160, 161, 165, 179, 191

microscope, 190

models, 6, 112, 138, 166, 168

momentum, 43, 57, 58, 59, 60, 61, 65, 75, 78, 81, 86, 87, 92, 93, 98, 103, 104, 105, 113

N

nanostructures, 17, 113

nanosystems, 2, 3, 4, 122

nanotube, vii, viii, 1, 2, 3, 4, 5, 6, 7, 8, 11, 12, 14, 15, 16, 18, 30, 34, 36, 38, 41, 42, 43, 44, 49, 54, 55, 56, 59, 75, 82, 83, 84, 86, 87, 91, 96, 101, 102, 109, 110, 111, 113, 114, 115

non-degenerate electron gas, 111, 130, 134, 135, 140, 141, 160, 173

normal state superconductor, 166, 169

nucleation, 4

Index

213

O

optical properties, 6, 204
optoelectronics, 188
orbit, 65, 81
oscillation, viii, 2, 11, 26, 43, 52, 71, 74, 112, 113
oscillators, 87
overlap, 32, 61, 78, 103, 104, 105, 106, 111
oxygen, 170

P

parallel, 5, 6, 7, 30, 43, 91, 163
particle physics, 119
periodicity, 146, 147, 155
permittivity, 5, 159, 160
phonon mediation factor, 148
phonons, 4, 148
photonics, 159
physics, viii, 2, 5, 115, 119, 121, 122, 123, 199
Planck constant, 18
plasma waves, viii, 2, 7, 85, 93, 98, 99, 100, 102, 109, 110, 113, 114
polarization, viii, 2, 7, 49, 93, 94, 96, 98, 99, 109, 114, 194
polarization operator, viii, 2, 7, 49, 93, 94, 96, 98, 99, 109, 114
probability, 126, 127, 128, 129, 130, 133, 200
probability density (function), 128, 129, 133
probability density function, 128, 129, 133
probability distribution, 128, 129

Q

quantization, vii, viii, 1, 2, 3, 16, 29, 63, 80, 91
quantum mechanics, viii, 3, 123, 145

quantum well, 3, 188, 190, 196, 200

R

radiation, 160
radius, viii, 2, 6, 7, 8, 14, 18, 25, 30, 32, 34, 42, 44, 49, 52, 57, 59, 71, 75, 84, 85, 92, 103, 110, 111, 140
recommendations, iv
relaxation, 125, 133, 136, 141, 143, 149, 153, 154, 155, 157, 158, 160, 164, 167, 169, 171, 178, 179
relaxation time, 125, 133, 136, 141, 143, 149, 153, 154, 155, 157, 158, 159, 160, 164, 167, 169, 171, 178, 179
relaxation times, 153, 157, 169
resistance, 146, 164, 167, 194, 197, 202
resistivity, ix, 124, 136, 138, 145, 149, 153, 155, 156, 157, 158, 165, 167, 176, 179, 184, 186, 189, 200
resistivity temperature dependence, 124
resolution, 125, 190
response, 3, 45, 55
room temperature, 126, 135, 138, 153, 175, 176, 177, 178, 179, 188, 192
root, viii, 2, 6, 9, 20, 31, 34, 37, 52, 71, 83, 92, 111

S

Samsung, 187, 207, 208
sapphire, 188, 190, 192
scattering, vii, ix, 3, 4, 114, 124, 126, 136, 140, 145, 146, 147, 148, 149, 150, 151, 155, 156, 161, 162, 165, 178, 193
scattering cross-section, 145, 147, 148, 149, 150, 151, 155, 156, 161, 162, 179
schema, 195, 201
scientific papers, 121, 122
semiconductor, viii, 2, 3, 4, 5, 6, 7, 13, 17, 18, 19, 21, 23, 27, 29, 30, 43, 49, 54, 58,

214 *Index*

59, 65, 85, 87, 102, 110, 113, 114, 175, 192

semiconductors, vii, viii, 123, 124, 134, 140, 142, 173, 188

shape, viii, 2, 7, 100, 110, 114, 190

showing, 194, 196, 201

silicon, 173, 179, 180, 182, 183, 184, 185

solution, 40, 87, 100, 105, 189

Sommerfeld's model, 125, 130, 136, 139, 142, 152, 154

spin, viii, 2, 4, 7, 9, 18, 26, 30, 31, 41, 43, 44, 49, 50, 52, 71, 83, 103, 104, 105, 106, 107, 108, 109, 110, 114, 163, 166, 173

state, vii, viii, 8, 13, 18, 31, 32, 33, 35, 37, 44, 57, 111, 123, 124, 125, 134, 135, 145, 158, 164, 166, 167, 169, 173, 179, 188, 193

statistics, ix, 3, 54, 121, 122, 124, 125, 130, 135, 145, 175, 178

stochastic description, viii, 123, 125, 178

storage, 3

stress, 194, 199, 203, 204

structure, vii, viii, ix, 2, 3, 5, 114, 146, 187, 190, 191, 192, 193, 195, 200, 201

superconductor, ix, 124, 145, 165, 166, 169, 172, 179, 183

superlattice, vii, viii, 1, 2, 4, 6, 7, 30, 31, 33, 36, 37, 40, 55, 56, 60, 74, 75, 78, 80, 82, 83, 84, 85, 86, 87, 91, 92, 96, 99, 101, 102, 103, 104, 109, 110, 111, 113, 114, 189

susceptibility, 104, 107, 108, 163, 164

T

techniques, 189

technology, 2, 173, 189

temperature, vii, ix, 3, 12, 14, 15, 17, 22, 23, 27, 28, 29, 34, 35, 36, 37, 38, 39, 42, 43, 47, 54, 58, 59, 77, 98, 111, 124, 126,

127, 129, 130, 132, 133, 138, 143, 145, 146, 147, 148, 149, 150, 151, 153, 155, 156, 157, 158, 159, 160, 163, 165, 166, 167, 169, 170, 171, 172, 179, 192

temperature dependence, 17, 27, 42, 124, 166, 173

thermodynamic functions, 2, 4, 7, 15, 42, 43

thermodynamics, 119, 121, 122

transition temperature, 165

transparency, 85, 98, 100, 112, 113, 114

transport, vii, viii, 123, 124, 126, 153, 158, 165, 173, 178

tunneling, vii, ix, 187, 188, 189, 196, 200, 203, 204

tunneling junction, 188, 189, 196, 200, 203, 204

V

vacancies, 146

valence, ix, 8, 124, 125, 128, 129, 130, 131, 133, 134, 142, 145, 174, 200

velocity, 48, 51, 86, 96, 102, 114, 133, 134, 136, 140, 143, 149, 151, 152, 153, 157, 160, 161, 162, 165, 169, 179, 188

vibration, 145, 147

W

wave number, 92, 93, 94

wave propagation, 112, 114

wave vector, 6, 8, 30, 45, 56, 85

weakness, 125

windows, 85, 98, 112, 113, 114

wires, 3, 4, 192

Y

YBCO, 165, 169, 170, 171, 172, 183

Related Nova Publications

Spintronics: A Review and Directions for Research

Editor: Arthur R. Hampton

Series: Physics Research and Technology

Book Description: This compilation first presents a brief literature overview of ferromagnetism in zinc oxide, as well as a survey on ion implantation and irradiation-mediated ferromagnetism. The authors highlight the intrinsic and extrinsic origins of ferromagnetism in 200 KeV Ni+2 ion implanted ZnO (Ni: ZnO)/undoped ZnO.

Softcover ISBN: 978-1-53614-526-7
Retail Price: $95

An Introduction to Molecular Dynamics

Editor: Mark S. Kemp

Series: Physics Research and Technology

Book Description: In the opening chapter of *An Introduction to Molecular Dynamics*, the method of statistical geometry, based on the construction of a Voronoi polyhedral, is applied to the pattern recognition of atomic environments and to the investigation of the local order in molecular dynamics-simulated materials.

Softcover ISBN: 978-1-53616-054-3
Retail Price: $95

To see a complete list of Nova publications, please visit our website at www.novapublishers.com

Related Nova Publications

Energetic Particles and Auroras in Magnetosphere/Ionosphere

Author: Lev I. Dorman

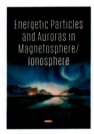

Series: Physics Research and Technology

Book Description: We hope that this review-book will be interesting and useful for researches, engineers, students of corresponding specialties, and all people interested in developing of modern technologies in space and in problems of Geomagnetosphere, Ionosphere, Upper and Low Atmosphere, Space Weather and Space Climate, and how they influence on the Earth's Civilization.

Hardcover ISBN: 978-1-53615-904-2
Retail Price: $310

A Closer Look at Biomechanics

Editor: Daniela Furst

Series: Physics Research and Technology

Book Description: The research presented in the opening chapter of *A Closer Look at Biomechanics* discusses the use of bone cements, and tests how a novel bone cement, medical grade two-component injectible polymer on silicone basis, can be used.

Softcover ISBN: 978-1-53615-866-3
Retail Price: $95

To see a complete list of Nova publications, please visit our website at www.novapublishers.com